ASHP Assisted Solar Energy Systems:
Classification, Applications, and Performance Research

空气源热泵辅助太阳能能源系统的分类、应用和性能研究

Wang Xinru   Cui Tong   Sun Tianmei   Wu Jinshun   Pan Song

中国建筑工业出版社
CHINA ARCHITECTURE & BUILDING PRESS

图书在版编目（CIP）数据

空气源热泵辅助太阳能能源系统的分类、应用和性能研究 = ASHP Assisted Solar Energy Systems: Classification, Applications, and Performance Research : 英文 / 王新如等著. -- 北京 : 中国建筑工业出版社, 2024.8. -- ISBN 978-7-112-30158-4

Ⅰ. TM615

中国国家版本馆 CIP 数据核字第 2024TP0408 号

责任编辑：率　琦　刘文昕
责任校对：赵　力

ASHP Assisted Solar Energy Systems:
Classification, Applications, and Performance Research
空气源热泵辅助太阳能能源系统的分类、应用和性能研究
Wang Xinru　Cui Tong　Sun Tianmei　Wu Jinshun　Pan Song

*

中国建筑工业出版社出版、发行（北京海淀三里河路9号）
各地新华书店、建筑书店经销
北京点击世代文化传媒有限公司制版
建工社（河北）印刷有限公司印刷

*

开本：880毫米×1230毫米　1/32　印张：4⅛　字数：140千字
2024年8月第一版　2024年8月第一次印刷
定价：49.00元
ISBN 978-7-112-30158-4
（43086）

版权所有　翻印必究
如有内容及印装质量问题，请与本社读者服务中心联系
电话：（010）58337283　QQ：2885381756
（地址：北京海淀三里河路9号中国建筑工业出版社604室　邮政编码：100037）

Preface

With the ever-growing global energy demand, the development and utilization of renewable energy sources have become an urgent priority. Solar energy, as an inexhaustible and clean green energy, possesses tremendous development potential. However, the intermittency and instability of solar energy limit its widespread application in practical use. The ASHP-assisted solar energy system, as a new type of integrated renewable energy utilization technology, combines solar energy with ground-source heat pumps, which can effectively improve energy efficiency and reduce energy consumption.

This book aims to systematically introduce the classification, applications, and performance research of ASHP-assisted solar energy systems. The book is divided into three parts: The first part mainly discusses the classification of ASHP-assisted solar energy systems, including classification methods based on different heat source and heat pump types, solar collector types, and other dimensions. This part will also explore the advantages and disadvantages of different types of systems to help readers better understand and choose the system that suits them. The second part focuses on the applications of ASHP-assisted solar energy systems, covering areas such as building heating, cooling, domestic hot water, and more. It will also analyze specific cases to demonstrate the effectiveness and advantages in real-world applications. The third part delves into the performance research of ASHP-assisted solar energy systems, including evaluation methods and technologies for system energy efficiency ratio, reliability, and economy, providing a theoretical basis for optimizing system design and improving operational efficiency.

This book is comprehensive, well-structured, and can serve as a reference for researchers in the field of renewable energy, as well as for students and engineers in the fields of energy engineering and building environment and energy application. We hope that this book will provide useful insights and guidance for the research and application of ASHP-assisted solar energy systems.

Finally, we would like to thank all the authors who participated in the writing of this book, whose hard work and professional contributions made the completion of this book possible. We also thank the staff of the publishing house for their efforts in bringing this book to publication. We hope that this book will contribute to the development and application of renewable energy, and help build a beautiful and harmonious society.

Editorial Board of *ASHP Assisted Solar Energy Systems: Classification, Applications, and Performance Research.*

Contents

Preface

Chapter 1 Introduction 1
   1.1 Introduction to this book 1
   1.2 The main content of this book 2
   1.3 Book outline 3
   References 5

Chapter 2 Introduction to the Solar Energy System 8
   2.1 Common solar thermal assisted ASHP (ST-ASHP), photovoltaic combined ASHP (PV-ASHP) and photovoltaic/thermal (PV/T-ASHP) 10
       2.1.1 System configuration 10
       2.1.2 ST-ASHP system 10
       2.1.3 PV-ASHP system 11
       2.1.4 PV/T-ASHP system 12
   2.2 Comparison of different solar assisted ASHP systems 13
   2.3 Summary 15
   References 16

Chapter 3 Research of ST-ASHP System 18
   3.1 Research orientation 18
   3.2 ST-ASHP system performance 26
   3.3 Comparison of ST-ASHP systems 30
   3.4 Summary 32
   References 33

Chapter 4 Research of PV-ASHP System 36
    4.1   Research orientation 36
    4.2   PV-ASHP system performance 37
    4.3   Summary 40
    References 41

Chapter 5 Research of PV/T-ASHP System 42
    5.1   Research orientation 42
    5.2   Experiment setting of PV/T-ASHP system 46
          5.2.1   Simulation models of the solar absorber 47
          5.2.2   Simulation models of the designed absorber 52
    References 58

Chapter 6 Experiment Study of PV/T-ASHP System in an Office Room Application 59
    6.1   Design of PV/T-ASHP system 59
          6.1.1   The detailed information about an office room 59
          6.1.2   Design of the system 61
          6.1.3   Measurement and calculation 65
          6.1.4   Conclusion 66
    6.2   Experiment of testing of the solar collector in PV/T-ASHP system and model validation using Matlab 66
          6.2.1   Working principle of the solar collector 68
          6.2.2   Matlab model of the solar collector 71
          6.2.3   Validation of the Matlab model 75
          6.2.4   Parametric study of the solar collector 76
          6.2.5   Comparison of the experiment and the Matlab model 78
          6.2.6   Conclusion 80
    6.3   PV/T-ASHP system performance 81
    6.4   Performance study of PV/T-ASHP system 83
          6.4.1   Description of the experimental device 83
          6.4.2   Working principle of PV/T-ASHP system 85
          6.4.3   Calculation of PV/T-ASHP system performance 85

	6.4.4	Analysis of the result	86
	6.4.5	Economic analysis	95
	6.4.6	Environmental analysis	96
6.5	Comparison of the proposed system with the previous research results		97
6.6	Summary		97
	References		99

Chapter 7　ANN Prediction Study 102
- 7.1 Introduction 102
  - 7.1.1 Background 102
  - 7.1.2 Data collecting 109
  - 7.1.3 Methodology 110
  - 7.1.4 Evaluation criteria 111
- 7.2 Data observation 112
- 7.3 ANN 115
- 7.4 Results 116
  - 7.4.1 ANN model with one hidden layer 116
  - 7.4.2 ANN model with two hidden layers 118
- 7.5 Results comparison 121
- 7.6 Summary 122
- References 123

Chapter 8　Conclusion and Future Work 127
- 8.1 Conclusion of the work 127
- 8.2 Future work 128
- References 129

Chapter 1
Introduction

1.1 Introduction to this book

Photovoltaic/thermal (PV/T) systems play an important role in solar system development, reducing pollution, and meeting the rapid increase in energy demand. PV/T can significantly improve energy transfer efficiency, hence improving its performance which has attracted extensive research attention. Different structures of the solar collectors are proposed. To reduce the investment and cost of the experiment, several prediction studies are also conducted, as modeling its performance is meaningful for the optimization and design of the system.

In this book, a photovoltaic/thermal assisted air source heat pump (PV/T-ASHP) system with a novel absorber of the solar collector in Beijing was studied. It is shown that for the whole year the electricity self-sufficiency is over 50% and the heat self-sufficiency is 30.7% in extreme conditions for the heating systems (e.g., December). Ambient temperature and solar irradiation changes highly affect the performance of the whole system. In extreme weather, e.g., on a sunny day in December, the average COP is higher than 3.5. The payback time for this system is 9.5 years, and it reduces equivalent to 6.8 t of carbon dioxide (CO_2) emission per year. Based on the experiment results, we then predict the performance using an ANN. It is seen that the more considered factor, the higher the performance of the prediction model. It is also seen that in this case, an ANN with two hidden layers performs better than that with one hidden layer. Furthermore, the comparison results suggest that in practice it is required to find the best suitable ANN network before application.

1.2 The main content of this book

Recently, the concerns about the environmental effects of energy consumption have been significantly increased. The growing concerns encourage the versatile development of renewable and clean energy systems which can reduce the environmental pollution caused by fossil fuels, e.g., oil, and coal.[1] And the global renewable energy power capacity increased rapidly from 2007 to 2017.[2] The total installed renewable power capacity at the end of 2017 is the highest in China, which is nearly 30% of the world's renewable power capacity—approximately 647 GW, including 313 GW of hydropower. Among all renewable and sustainable energies, such as wind energy, soil energy, nuclear energy, solar energy is often considered the most important due to its accessibility throughout the world.[3] The renewable power capacity in China accouted for 75% of global capacity in 2017 and China continues to be the leader in solar-based energy production in the world. Solar photovoltaic (PV) is the top source of new power capacity in several major markets, including China, and India.

Renewable energy systems based on solar energy are based on either solar-thermal energy or PV. Solar thermal converts solar energy into heat whereas PV converts solar energy into electric energy.[4] However, in the PV system, besides the electricity produced by the cells, the solar energy also increases surface temperature of the cells. This may reduce the cells' efficiency in producing electric energy. Therefore, the heat produced by PV cells needs to be managed to reduce the surface temperature and improve the performance of the cells. To address this issue, one approach is to combine PV with the solar thermal system, i.e., PV/T systems. PV/T systems enable dual usage of solar energy and produce both electricity and heat. PV/T improves the efficiency of PV cells, and further produces more energy than either solar thermal or PV system.[5]

There are several different designs for thermal absorbers. The common designs include a sheet-and-tube structure, a rectangular tunnel with or without fins/grooves, flat plate tube, micro-channel heat pipe array/ heat mat, extruded heat exchanger, roll-bond heat exchanger, and cotton wick structure. [6] The sheet-and-tube absorber needs a lower investment due to the established industry and also provides high efficiency.[7-8] However, these absorbers are complex [9], and their heavy weights limit their applications.[10] Compared with other absorbers, the rectangular tunnel with or without fins/grooves has a lower weight and requires less investment[11-12], but the heat transfer efficiency is also lower. [13-17] For a flat plate tube, the resistance is high, although the contact between PV cells and absorber is simple. [18-19] The micro-channel heat pipe, also called heat mat, improves the heat transfer performance, but the heat transfer resistance is also increased. [20-22] Although the extruded heat exchanger requires a special design, its performance is high. [23] The roll-bond heat exchanger and cotton wick structure absorbers are limited in the application. [24-28] Currently, extruded heat exchanger is becoming a promising solution. Different arrangements and combinations can also create various flow channels. [23] However, only a few related studies are investigating the performance of such absorbers. Xu et al. [23] Developed a super thin-conductive thermal absorber to regulate PV working temperature by retrofitting the existing PV panel into PV/T panel. The schematic structure of the thermal absorber and the associated prototypes were also shown in. [23] They found that the hybrid PV/T panel enhances the electrical return of PV panels by nearly 3.5% and increases the overall energy output by nearly 324.3%. The laboratory testing results demonstrated that PV/T panel achieves an electrical efficiency of about 16.8% (relatively a 5% improvement comparing to the stand-alone PV panel), and yields an extra amount of heat with a thermal efficiency of nearly 65%. Shen et al. [29] Proposed a novel compact solar thermal façade (STF) with an internally extruded pin-fin flow channel. They showed that this construction of STF is beneficial for further design, optimization, and application including the provision of hot water, space heating/cooling, increased ventilation, or even electricity in the buildings.

1.3 Book outline

This book contains seven chapters: Chapter 2 presents the literature review of the existing related researches on solar assisted air source heat pump system. Chapter 3 presents ST-ASHP system. Chapter 4 presents PV-

ASHP system. Chapter 5 presents PV/T-ASHP system. Chapter 6 is the experimental results of this system. Chapter 7 presents the prediction of the performance of the PV/T collector using an artificial neural network (ANN). And Chapter 8 presents the summaries. The main results of case study of PV/T-ASHP system are concluded and summarized in this chapter. Besides, a few future research directions are described.

References

[1] Ellabban O, Abu-Rub H, Blaabjerg F. Renewable energy resources: Current status, future prospects and their enabling technology[J]. Renewable and Sustainable Energy Reviews, 2014, 39: 748-764.

[2] Raturi A K. Renewable 2018 Global Status[J]. 2018.

[3] Shahsavar A, Ameri M. Experimental investigation and modeling of a direct-coupled PV/T air collector[J]. Solar Energy, 2010, 84(11): 1938-1958.

[4] Baljit S S S, Chan H Y, Sopian K. Review of building integrated applications of photovoltaic and solar thermal systems[J]. Journal of Cleaner Production, 2016, 137: 677-689.

[5] TIEA (2010), Technology Roadmap - Solar Photovoltaic Energy 2010[R], IEA, Paris https://www.iea.org/reports/technology-roadmap-solar-photovoltaic-energy-2010, Licence: CC BY 4.0.

[6] Wu J, Zhang X, Shen J, et al. A review of thermal absorbers and their integration methods for the combined solar photovoltaic/thermal (PV/T) modules[J]. Renewable and Sustainable Energy Reviews, 2017, 75: 839-854.

[7] He W, Zhang Y, Ji J. Comparative experiment study on photovoltaic and thermal solar system under natural circulation of water[J]. Applied Thermal Engineering, 2011, 31(16): 3369-3376.

[8] Charalambous P G, Maidment G G, Kalogirou S A, et al. Photovoltaic thermal (PV/T) collectors: A review[J]. Applied Thermal Engineering, 2007, 27(2-3): 275-286.

[9] Buker M S, Mempouo B, Riffat S B. Performance evaluation and techno-economic analysis of a novel building integrated PV/T roof collector: An experimental validation[J]. Energy and Buildings, 2014, 76: 164-175.

[10] Newform Energy. 2016. PV/T Water Collector [EB/OL]. Available: http://www.newformenergy.com [2024-04-17].

[11] Hassani S, Taylor R A, Mekhilef S, et al. A cascade nanofluid-based PV/T system with optimized optical and thermal properties[J]. Energy, 2016, 112: 963-975.

[12] Zhang X, Zhao X, Smith S, et al. Review of R&D progress and practical application of the solar photovoltaic/thermal (PV/T) technologies[J]. Renewable and Sustainable Energy Reviews, 2012, 16(1): 599-617.

[13] Zondag H A, De Vries D W, Van Helden W G J, et al. The yield of different combined PV-thermal collector designs[J]. Solar Energy, 2003, 74(3): 253-269.

[14] Ibrahim A, Othman M Y, Ruslan M H, et al. Recent advances in flat plate photovoltaic/thermal (PV/T) solar collectors[J]. Renewable and Sustainable Energy Reviews, 2011, 15(1): 352-365.

[15] Kroiß A, Präbst A, Hamberger S, et al. Development of a seawater-proof hybrid photovoltaic/thermal (PV/T) solar collector[J]. Energy Procedia, 2014, 52: 93-103.

[16] Chow T T, He W, Ji J. Hybrid photovoltaic-thermosyphon water heating system for residential application[J]. Solar Energy, 2006, 80(3): 298-306.

[17] Farshchimonfared M, Bilbao J I, Sproul A B. Full optimisation and sensitivity analysis of a photovoltaic–thermal (PV/T) air system linked to a typical residential building[J]. Solar Energy, 2016, 136: 15-22.

[18] Ibrahim A, Fudholi A, Sopian K, et al. Efficiencies and improvement potential of building integrated photovoltaic thermal (BIPVT) system[J]. Energy Conversion and Management, 2014, 77: 527-534.

[19] FotoTherm. 2016. FOTO-THERM Thermo-photovoltaic modules [EB/OL]. Available: https://www.fototherm.com/en/ [2024-04-17].

[20] Hou L, Quan Z, Zhao Y, et al. An experimental and simulative study on a novel photovoltaic-thermal collector with micro heat pipe array (MHPA-PV/T)[J]. Energy and Buildings, 2016, 124: 60-69.

[21] Zhou J, Zhao X, Ma X, et al. Experimental investigation of a solar driven direct-expansion heat pump system employing the novel PV/micro-channels-evaporator modules[J]. Applied Energy, 2016, 178: 484-495.

[22] Jouhara H, Szulgowska-Zgrzywa M, Sayegh M A, et al. The performance of a heat pipe based solar PV/T roof collector and its potential contribution in district heating applications[J]. Energy, 2017, 136: 117-125.

[23] Xu P, Zhang X, Shen J, et al. Parallel experimental study of a novel super-thin thermal absorber based photovoltaic/thermal (PV/T) system against

conventional photovoltaic (PV) system[J]. Energy Reports, 2015, 1: 30-35.

[24] Dupeyrat P, Ménézo C, Wirth H, et al. Improvement of PV module optical properties for PV-thermal hybrid collector application[J]. Solar Energy Materials and Solar Cells, 2011, 95(8): 2028-2036.

[25] Dupeyrat P, Ménézo C, Rommel M, et al. Efficient single glazed flat plate photovoltaic–thermal hybrid collector for domestic hot water system[J]. Solar Energy, 2011, 85(7): 1457-1468.

[26] Bai Y, Chow T T, Menezo C, et al. Analysis of a hybrid PV/thermal solar-assisted heat pump system for sports center water heating application[J]. International Journal of Photoenergy, 2012, 2012.

[27] Bionicol. 2010. Bionicol-Dev a bionic Sol Collect Alum Roll absorber-Proj Status Second year [EB/OL]. Available: http://www.bionicol.eu/ [2024-04-17].

[28] Chandrasekar M, Senthilkumar T. Experimental demonstration of enhanced solar energy utilization in flat PV (photovoltaic) modules cooled by heat spreaders in conjunction with cotton wick structures[J]. Energy, 2015, 90: 1401-1410.

[29] Shen J, Zhang X, Yang T, et al. Characteristic study of a novel compact Solar Thermal Facade (STF) with internally extruded pin–fin flow channel for building integration[J]. Applied Energy, 2016, 168: 48-64.

Chapter 2
Introduction to the Solar Energy System

Solar energy and heat pump systems have been investigated for several decades, which are capable of increasing the share of renewable energy. Many articles have been published on these two topics. However, literatures on solar assisted air source heat pump (ASHP) system of different technologies (solar thermal, photovoltaic and hybrid thermal/photovoltaic) are lacking, and thus leading to a practical challenge to evaluate different solar assisted ASHP systems in various scenarios. [1] The ASHP systems assisted by solar energy resources have drawn wide attention owing to their great feasibility in buildings for space heating/cooling and hot water purposes. This review thus conducts a comprehensive review of the prevailing solar assisted ASHP systems, including their system boundaries, system configurations, performance indicators, research methodologies and major research findings.

We will concentrate on the R&D (research and development) work of three most prevailing solar assisted ASHP systems, including solar thermal assisted ASHP (ST-ASHP), photovoltaic combined ASHP (PV-ASHP) and photovoltaic/thermal (PV/T-ASHP).

System boundaries are important to understand the system and evaluate its performance in an equivalent condition. While in many studies, researchers do not clearly define the system boundaries of solar assisted ASHP systems, which makes it difficult to distinguish the different systems. Within IEA SHC Task 44/HPP Annex 38 (T44A38), the various system boundaries of solar and heat pump systems were introduced [2], shown in Figure 2-1. Heat sources are identified in green, while purchased energy electricity is in grey. The red boxes on the right represent the energy supplied to users.

Chapter 2
Introduction to the Solar Energy System

For conventional ASHP system, the heat pump gains heat from air and electricity from the grid. In theoretical working principle, conventional ASHP system can supply heating, cooling and hot water for users, but with the climate limitations, ASHP systems generally need auxiliary heaters to meet the users' demand in severe conditions. Solar assisted ASHP system can use both air and solar energy as the energy sources. The system configuration is explained in section 2.2 and the definitions of system boundaries are shown in Figure 2-1. There are also other sources of heat pump system, like ground, water and so on. In this book, we only show the air source heat pump system.

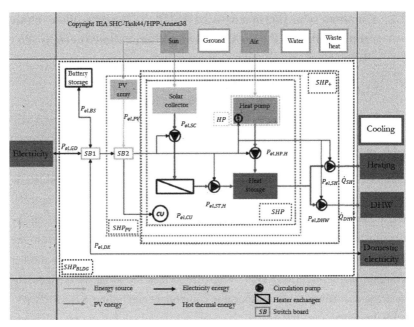

Figure 2-1 Different system boundaries (the figure is based on the ongoing Task 53 [2] work)

According to the Task 53 [3], the definitions of solar assisted air source heat pump system boundaries (Figure 2-1) mainly contain five types. The basic air source heat pump system (HP) consists only of the heat pump unit, outlined by orange dashed lines. Adding solar collector system, which is concluded in green dashed line, the system is called solar assisted air source heat pump system (SHP), while the above two system types are both without circulation water pump. System boundary SHP+ does include these pumps (blue dashed line). Solar assisted air source heat pump system with PV arrays (SHP_{PV}) is shown in red dashed line. It includes the PV array but without battery storage. With the battery storage system, the system boundary is called solar assisted air source heat pump system (SHP_{BLDG}) for building heating, DHW and electricity (purple dashed line).[4] Corresponding to Task 53, in this book, SHP + , SHP_{PV} boundary and SHP_{BLDG} are recorded as ST-ASHP system, PV-ASHP system and PV/T-ASHP system respectively.

2.1 Common solar thermal assisted ASHP (ST-ASHP), photovoltaic combined ASHP (PV-ASHP) and photovoltaic/thermal (PV/T-ASHP)

2.1.1 System configuration

There are many different ways of categorization of solar assisted ASHP system. In Figure 2-1, the system boundaries are identified on a component level, e.g., heat pump, or on a system level. Among them, the solar heat pump (SHP+) system includes the circulation pumps, while SHP and HP do not, which is used for comparing the assessment of the system environmental impact in operation.[2] Besides that, they can also be divided based on the type of heat demand to be served, e.g., heating, DHW.[5] In addition, the interaction way of the heat pump and the solar assisted system is another common categorization. This book categorizes the solar assisted ASHP systems according to the evaporator component, including ST-ASHP, PV-ASHP, and PV/T-ASHP.

2.1.2 ST-ASHP system

For ST-ASHP system, the solar thermal collector converts solar irradiation into thermal energy that could be used directly by users or supplied to ASHP. ASHP produces heat/cold for users. ST-ASHP system could supply different types of energy such as heating, cooling, hot water, even the steam.

Based on the generation method, ST-ASHP system can be categorized into direct expansion solar assisted system (DX-SA) and indirect expansion solar assisted system (IDX-SA). In DX-ST-ASHP, the heat pump and solar thermal system work together as one combined system and solar thermal collector acts as the system evaporator. Indirect expansion solar thermal and air source heat pump (IDX-ST-ASHP) systems can generally be classified in parallel, serial, regenerative (Figure 2-2) and complex system concepts by the interaction between solar thermal and heat pump. [2, 6] The serial type is common in solar and ground source heat pump combination system, while for solar thermal assisted air source heat pumps system, a statistical analysis on market-available ST-ASHP systems can be found in Ruschenburg et al. [7-8] Their main result indicates that the parallel systems are the market-dominating system concepts (61% of the surveyed systems). [8] In parallel configuration [Figure 2-2 (a)], solar collector and air source heat pump independently supply heating and/or DHW, or cooling to users.

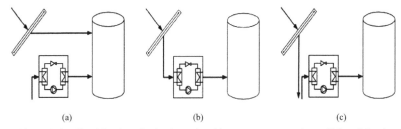

Figure 2-2 Classification of solar thermal and heat pump systems in parallel, serial and regenerative system concepts: (a) parallel; (b) serial; (c) regenerative [5-6]

ST-ASHP system commonly use two types of media: liquid and air. Based on media temperature, ST-ASHP is divided into two types. Low and medium temperature is less than 100℃ .[7] The common temperature used in real application of ST-ASHP system is usually low and medium.

2.1.3 PV-ASHP system

Solar photovoltaic (PV) [9] can directly convert sunlight into electricity without any heat engine to interfere. The ASHP partially covered by the electricity (produced by PV) in the buildings is called PV-ASHP system. Nowadays, another popular type of PV integrated with the building such as roofing, windows and so on is called building-integrated photovoltaic (BIPV) systems. PV-ASHP system could supply heating, cooling and electricity to buildings. The principle of PV/T-ASHP system is that the electricity produced by PV can be supplied to ASHP which supplies the heating or cooling to the building. When PV cannot meet the demand, ASHP can gain electricity from the grid. Meanwhile, the surplus electricity produced by PV system can be fed back into the grid, e.g., midday on sunny days when the load of building is also small.

PV-ASHP system could be classified as direct expansion and indirect expansion types. Moreover, it can also be classified by the array types [10], while the common classification is based on the lighting absorbing materials. The most common material is silicon which includes silicon, amorphous silicon and crystalline silicon. There are other materials such as cadmium telluride and cadmium sulphide, organic and polymer cells and hybrid photovoltaic cell as well as some other cells.

2.1.4 PV/T-ASHP system

Normally, PV module only converts 4%-17% of incoming solar radiant into electricity, which varies along with the material and working conditions, and as a result, much more solar energy is converted into heat. Over high ambient temperature could reduce the efficiency of PV and cause damage to the structure to the modules. [7] PV/T can produce electricity as well as heat, which can cool the surface temperature of cells. Like BIPV system, PV/T integrated with building is called BIPV/T system, which can save the space more than two separated arrays of PV and solar thermal collectors. Combined with ASHP, PV/T-ASHP system can supply heating and domestic hot water (DHW), even cooling for buildings in the future.

While in PV/T-ASHP system, as PV/T collectors can not only provide heat, but also produce electricity, the common type is in serial connection, as PV/T collectors are used to regenerate the heat exchanger. [11] PV/T systems can be divided into five types: air cooled system, water cooled system, both water and air cooled system, using PCM PV/T system and using heat pipes in PV/T system. [12] And more recently, using PCM in PV/T system is concentrated. [13] And based on the integrated fabric of the building, BIPV/T system can be divided into wall integrated, window integrated, roof integrated and façade integrated systems. For the whole PV/T-ASHP system, according to the application, there are heating, DHW intents types.

2.2 Comparison of different solar assisted ASHP systems

For conventional ASHP system, compared with other source heat pump systems, such as water, ground and so on, the main advantages of ASHP system conclude that the investment is much lower and the installation is much easier than other ground source heat pump systems, and the operation and management of ASHP system are also easier. However, the disadvantages are also obvious, the performance of ASHP systems is significantly influenced by ambient temperature. Water and ground source heat pump systems are more stable and the performance of them is higher than ASHP system, especially under extreme conditions, the frosting is still one of the big problems for ASHP system, which seriously affects the performance of ASHP. Auxiliary heat will increase the extra electricity energy consumption.

For solar assisted air source heat pump system, including ST-ASHP, PV-ASHP and PV/T-ASHP system, compared with ASHP system, the common advantages for them are obvious. From the system aspect, performance of them is better than ASHP system [14], as the solar assisted systems gain energy from both air and solar, which are both renewable and sustainable. In above reviewed part, the COP of the solar-assisted air source heat pump system increased by 30% to 60% compared to using ASHP system alone, with higher values under certain specific conditions. As ASHP system is heated by solar energy, the frosting issue is also reduced. As a result, the solar assisted air source heat pump system has a wider spread than ASHP system, especially for cold and more extreme regions. [15] Moreover, for the whole energy aspect, the solar assisted air source heat pump system could supply more energy to users and produce more types of energy, i.e., PV-ASHP and

PV/T-ASHP system can supply electricity for buildings. Solar assisted air source heat pump system could meet the increasing energy demand better than ASHP system. Besides that, as mentioned above, the solar and air energy are clean and friendly to environment, the solar assisted systems could reduce the environment pollution caused by fossil fuels, i.e., coil, oil. Along these obvious advantages, it is certainly that there are some common weaknesses for all solar assisted air source heat pump systems. Comparing with ASHP system, the investment concluding equipment, installation and maintenance is much higher and the operation and management are much more complex, as the systems are complex.

While only for solar assisted air source heat pump systems, ST-ASHP, PV-ASHP and PV/T-ASHP systems, there are also some comparison results of their benefits. Among solar assisted air source heat pump systems, ST-ASHP can avoid frosting in winter, especially in cold regions. Even below ambient temperature, the system performance is still relatedly high. Moreover, the installation requirement is much less strict than PV-ASHP and PV/T-ASHP systems, causing that the payback time is the shortest among three solar assisted air source heat pump system. However, the disadvantage is that most energy produced by ST-ASHP is used only for heat or hot water, and ST-ASHP system is still affected more by the external weather conditions than other two systems, as ST-ASHP system is significantly related with ambient temperature, while other two solar assisted air source heat pump systems mainly rely on the solar radiation. In terms of PV-ASHP system, this system is usually parallel, the PV cell produce electricity supplied to equipment consumption or directly to buildings, while ASHP system supply the heat and hot water (with water tank system) to users. Besides the heat energy, electricity energy can be produced for direct use, which is improved than ST-ASHP system. The performance of the air source heat pump is the best among three solar assisted air source heat pump system, as the energy input is much lower. Of course, more complex system control strategy than ST-ASHP system is also required among PV generation, grid import, battery or TES unit. And electricity storage is still one of difficulties which hinder the widespread use. As for the most complex system among solar assisted air source heat pump systems, PV/T-ASHP system has its own characteristics. It can produce different types of energy including heat, hot water and electricity. Similar with PV-ASHP, but not the same, it can produce more capacity of energy and the users consume the lowest energy among three solar assisted air source heat pump systems, as this system has the highest and maximized solar utilization. But at this step, there are also some problems faced by PV/T-ASHP system. The most complex control system is basic for operation and management. The maintenance is much difficult. All these cause that it has the longest payback time. These characteristics restrict the development of PV/T-ASHP systems. [16] The summary of the main characters of solar assisted air source heat pump systems is given in Table 2-1.

Table 2-1 Summary of comparison among solar energy assisted ASHP systems

System	Advantage	Disadvantage
ST-ASHP	High efficiency; Shortest payback time; Low installation requirement; Lowest investment	Affected by ambient temperature
PV-ASHP	Saving electricity, even can support other equipments; Highest COP of HP; Lower input power	More investment; Existing electricity storage problem; Affected by solar irradiation; Installation problem
PV/T-ASHP	Highest solar energy utilization; Supply heat and electricity at the same time; Most energy production and lowest consumption	Most investment; Most complex system and control; Most installation and maintenance problem; Longest payback time

2.3 Summary

In this chapter, the definition of different solar energy systems are introduced firstly. Then, three common solar energy systems are presented including ST-ASHP, PV-ASPT and PV/T-ASHP. Finally, these three systems are compared based on existing researches.

References

[1] REN21. (2018). Renewables 2018 Global Status Report[R]. REN21 Secretariat. https://www.ren21.net/gsr-2018/.

[2] Solar and heat pump systems for residential buildings[M]. John Wiley & Sons, 2015.

[3] International Energy Agency Solar Heating and Cooling Programme (IEA SHC). (2017). New Generation Solar Cooling & Heating Systems (PV or solar thermally driven systems)[R]. Retrieved May 23, 2017, from http://task53.iea-shc.org.

[4] Poppi S, Sommerfeldt N, Bales C, et al. Techno-economic review of solar heat pump systems for residential heating applications[J]. Renewable and Sustainable Energy Reviews, 2018, 81: 22-32.

[5] Frank, E., Michel, H., Herkel, S., & Jorn, R. (2010). Systematic classification of combined solar thermal and heat pump systems[C]. Proceedings of the EuroSun 2010 Conference, Graz, Austria.

[6] Frank, E., Haller, M. Y., Herkel, S., & Ruschenburg, J. (2010). Systematic classification of combined solar thermal and heat pump systems[C]. Proceedings of the EuroSun 2010 Conference, Graz, Austria.

[7] Ruschenburg J, Herkel S, Henning H M. A statistical analysis on market-available solar thermal heat pump systems[J]. Solar Energy, 2013, 95: 79-89.

[8] Ruschenburg J, D'Antoni S E, Ellehauge K, et al. A review of market-available solar thermal heat pump systems[J]. A Technical Report of Subtask A, 2013.

[9] World Energy Council[R]. (2007). Survey of Energy Resources. World Energy Council.

[10] Rauschenbach H S. Solar cell array design handbook: the principles and technology of photovoltaic energy conversion[M]. Springer Science & Business Media, 2012.

[11] Joyce A, Coelho L, Martins J F, et al. A PV/T and heat pump based trigeneration system model for residential applications[C]//ISES Solar World Congress. 2011.

[12] Al-Waeli A H A, Sopian K, Kazem H A, et al. Photovoltaic/Thermal (PV/T) systems: Status and future prospects[J]. Renewable and Sustainable Energy Reviews, 2017, 77: 109-130.

[13] Qiu Z, Ma X, Zhao X, et al. Experimental investigation of the energy performance of a novel Micro-encapsulated Phase Change Material (MPCM) slurry based PV/T system[J]. Applied energy, 2016, 165: 260-271.

[14] Good C. Environmental impact assessments of hybrid photovoltaic–thermal (PV/T) systems–A review[J]. Renewable and Sustainable Energy Reviews, 2016, 55: 234-239.

[15] Zhou J, Zhao X, Ma X, et al. Experimental investigation of a solar driven direct-expansion heat pump system employing the novel PV/micro-channels-evaporator modules[J]. Applied Energy, 2016, 178: 484-495.

[16] Good C, Andresen I, Hestnes A G. Solar energy for net zero energy buildings–A comparison between solar thermal, PV and photovoltaic–thermal (PV/T) systems[J]. Solar Energy, 2015, 122: 986-996.

Chapter 3
Research of ST-ASHP System

3.1 Research orientation

In terms of ST-ASHP system, researchers use simulation aim to optimize the system, such as the structure, operation and efficiency, or to verify the performance of the system, as some experiments can not achieve in real projects. Weather data mainly containing temperature, solar radiation and seasonal change are the most important input parameters for simulation. The output parameters contain SPF, COP and tank volume and so on.

Most of the simulation studies focus on the optimization performance of ST-ASHP system. Many researches have compared their own proposed system with original ASHP system. For example, Jonas et al. [1] developed and validated a tool "SHP-SimFrame" in TRNSYS to simulate the solar heat pump system (SHP), especially solar thermal heat pump. Kim et al. [2] carried out three indirect solar assisted heat pump systems including the serial type solar assisted heat pump, solar assisted heat pump with hybrid solar collectors and parallel solar assisted heat pump system with flat plate solar thermal collectors. Zhang et al. [3] studied the structural parameters on the performance of a direct expansion solar assisted heat pump including the solar collector area, the collector thickness and the pipe length and internal diameter of condenser. And they got the optimized structural parameters for their system.

To make table 3-1 clearer, we chose several researches including the media studies except R22 and a few typical studies with new equipment. Almost all the simulation papers through software are conducted in TRNSYS software. i.e., Chen et al. [4] used TRNSYS to optimize ST-ASHP using CO_2 as heat pump refrigerant. The simulated results showed the optimized system has a high performance with the solar fraction can reach 71.1%, except Jonas et al. [1] They developed and validated a tool "SHP-SimFrame" in TRNSYS to simulate the solar heat pump system (SHP), especially solar thermal heat pump. They also only used the simulation method to compare the performance of ST-ASHP system against conventional ASHP system under various boundary conditions, but the conclusion was not representative. As the simulation parameters changed, the conclusion would also change.

Besides the simulation software method, Li et al. [5] and Deng et al. [6] developed their own mathematic models to simulate the performance of ST-ASHP systems, respectively. Both of them calculated the COP value and found that solar irradiation and ambient air temperature have a great effect on the system performance. They indicated that the solar efficiency decreases with the growth of solar irradiation and decrease of ambient temperature. For detail information, Li et al. [7] set up the parameters for components in TRNSYS and calculated SPF and SF of the system. It is proved that the overall efficiency of a DHW system was improved by the combination of a solar collector and heat pump. Deng et al. [6] proposed a modified direct expansion solar assisted air source heat pump water heater system, which is illustrated in Figure 3-1. The proposed system mainly contains a flat-plate solar collector, an evaporator, a compressor, a hot water tank with heat exchanger as a condenser, receiver and two electronic expansion valves (EEV). There are two operation modes. When the solar radiant is high enough to heat the water efficiently, it runs on single solar collector mode. While during the solar radiant is unavailable or low, and evaporating temperature is much lower than ambient air, it works under combination mode. As a result, at solar radiation of 100 W/m^2, the heating time of this system decreases by 19.8% compared to renovation when water temperature reaches 55℃. Meanwhile, the average system COP increases by 14.1%.

While all the results got by simulation were optimized, the performance of these studies were simulated under their defined boundary conditions. Among them, Fraga et al. [8] not only simulated the achievable value of SPF, but also calculated the potential SPF value. They developed a simulation

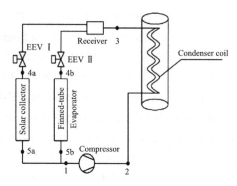

Figure 3-1 Schematic diagram of the new system [6]

model to analyse the potential performance of a ST-ASHP system. SPF of system achieved 4.4 and when temperature distributed SPF was from 3.1 to 4.1. Moreover, SPF of system achieved 5 is potential, but need considering the envelope, SH distribution temperature and solar collector area. In real project, the system should take account of real conditions, e.g., investment and roof area.

Among all the researches of solar assisted ASHP system, there is one special condition that the performance researches under extreme conditions. The common problem under extreme conditions is that ASHP would be frosting, which has also gained many concentrations in recent decades. [9-10] There are few researches focusing on ST-ASHP system that operates under extreme condition.

Qiu et al. [11] compared the integrated heating system of solar energy and air source heat pump system under different working conditions in cold regions. As a result, compared with high temperature and low temperature heating collecting systems, medium temperature has the best performance of this integrated system. Its COP is 55% higher than other two types when the outdoor temperature is −25°C.

Chapter 3
Research of ST-ASHP System

Table 3-1 Summary of simulation methods and the related results of ST-ASHP systems

Reference	System	Tool	Model	Location	Operation Conditions	Input data	Building Types	Evaluation parameters	Results
[1]	parallel ST-ASHP/ ST-GSHP	TRNSYS	Type 56-building; Type 832-collector; Type 401-compressor HP; Type 557a-borehole heat exchanger; Type 340-multiport storage tank	Helsinki, Strasbourg, Athens	solar heating/ CO_2 heat pump heating mode	Location	building type SFH45 (TRNSYS)	SPF	•The design of SHP systems depending on the boundary conditions of the specific application; •COP of ASHP= 3.66
[4]	solar combined CO_2 heat pump	TRNSYS	Type 71-solar collector; Type 841-CO_2 heat pump; Type 4- storage tank; Type 534- operation tank	Shanghai, China	—	Taiyuan, Shanghai	residential building	volume of storage tank and operation tank	•the optimization values: storage tank volume=2.21 m^3 and operation tank volume= 0.3 m^3; •the optimized system can save 14.2% electricity and solar fraction can reach 71.1%
[5]	ST-ASHP	TRNSYS 17	Type 1-quadratic efficiency collector; Type 3- pump; Type 2b-controller; Type 15-weather data reading and processing	Taipei and Kaohsiung, China	—	setting parameters for components	lab-scale system	SPF and solar fraction; Payback time	•The overall efficiency of a DHW system has improved, while the cost is also increased; • EPBT= 5 years; • SPF= 4.56 / 4.93 in Taipei and Kaohsiung

21

ASHP Assisted Solar Energy Systems:
Classification, Applications, and Performance Research

Continued

Reference	System	Tool	Model	Location	Operation Conditions	Input data	Building Types	Evaluation parameters	Results
[6]	DX-SHPWH	mathematical model	—	Xi'an, China	the solar radiation ranges; ambient air temperature; superheating degrees of the two evaporators	component parameters	—	heating capacity; COP	•Solar irradiation and ambient air temperature have a great effect on the system's performance; •The COP on average increases by 14.1%; •Given solar radiation of 500 W/m², the COP=4.98 and Qc = 605 W
[8]	ST-ASHP	TRNSYS	one-node model	Geneva, Switzerland	(i) direct solar heat production; (ii) storage discharge; (iii) activation of the HP, with surplus production; (iv) direct electric heating	hourly weather data; hourly load demand	multifamily buildings	SPF	•SPF of system achieved maximizes 4.4 and when temperature is distributed, the SPF is 3.1-4.1; •SPF of system achieves 5 is potential; •Investment and roof area condition are not considered

Continued

Reference	System	Tool	Model	Location	Operation Conditions	Input data	Building Types	Evaluation parameters	Results
[25] [26]	parallel ST-ASHP	TRNSYS 17	Type 832 QDT multimode model for collector; Type 340 for store; Type 805- heat exchanger; Type 887-heat pump	Zurich and Carcassonne, Sweden	SFH100 (100 single family houses) in Carcassonne	location and house model; interest rate; inflation rate and price of electricity	laboratory	SPF; Total electricity use; Annual DHW discharges energy; collector area; tank volume; UA-value of DHW heat exchanger; Payback time	• SPF= 2.84; • EPBT= 10 years; • system electricity changes between 305 kWh and 552 kW h/y when collector area from 5 m^2 to 15 m^2; • main influence for the system including annual DHW discharge energy, collector area, UA-value of DHW heat exchanger and heat pump size

Table 3-2 Summary of experiment methods and the related results of ST-ASHP systems

Reference	System	Medium	Operation mode	Location	Measured parameters	Building types	Evaluation parameters	Main results
[15]	ST-ASHP	R407c	heating mode	Taiyuan, China	environment condition	experiment room	energy transfer; COP; energy and exergy efficiency; process quality number; improvement potential	• Average energy efficiency of SIASHP is 77.67, 8% higher than ASHP and the average COP is 2.94, 8.1% more than ASHP; • Average process quality number is 59.36, 5.4% higher than ASHP
[16]	ST-ASHP	R407c	heating mode	Taiyuan, China	condensing temperature	experiment room	integrated part load value (IPLV); seasonal part load value (SPLV); relative size IPLV and SPLV	• The higher part load rate, the higher COP; • SPLV is no more than IPLV; • Average IPLV of SIASHP is 2.54 and SPLV is 2.53, 14.9% and 15.5% higher than ASHP respectively
[4]	solar combined CO_2 heat pump	—	solar heating/ CO_2 heat pump heating mode	Shanghai, China	ambient conditions	residential building	COP; total heat gain; solar fraction (SC) and SC efficiency	• System COPs of the solar heating and CO_2 HP heating modes can reach 13.5 and 2.18; • The studied solar assisted system can save 53.6% of the electricity consumption than CO_2 HP
[14]	Serial SA-ASHP	—	—	Erzurum, Turkey	weather data	residential building	COP; environmental and economic benefits; energy saving ratio; CO_2 reduction ratio; payback period	• The average COP of heat pump is 3.8 and solar assisted ASHP is 2.9

Chapter 3
Research of ST-ASHP System

Continued

Reference	System	Medium	Operation mode	Location	Measured parameters	Building types	Evaluation parameters	Main results
[17]	DMHP	—	heating/cooling mode	—	ambient condition, temperature/pressure of refrigerant	enthalpy difference lab	power consumption and COP; exergy analysis; heating capacity and power consumption; water temperature in SWT and DWT	• the COP is significantly affected by ambient temperature; • Heating capacity and COP of air source space heating mode are higher than that of solar space heating mode with the outdoor ambient temperature above 4℃; • Different components cause most exergy loss in different operation mode
[8, 27]	ST-ASHP	—	space heating / DHW	Geneva, Switzerland	weather data	Multifamily house	SPF; COP; energy flows; thermal storage	• The measured seasonal performance factor of the system is 2.9 for 2012; • Several points may have contributed to a relatively low SPF: (i) an unusually low demand for space heating along with an unusually high demand for domestic hot water; (ii) in absence of a load adjusted heat pump; (iii) a single heat distribution circuit with decentralized domestic hot water storage; (iv) no insulation of the unglazed solar collectors
[28-29]	ST-ASHP combined with PCM	—	heating/cooling/ water from freezing	Shenyang, China	ambient temperature; solar hot water temperature	laboratory	COP; Capacity	• Ambient temperature had a significant influence on the system's performance in cooling mode; • In heat pump cycle, solar-heated water temperature has a significant effect on system performance

3.2 ST-ASHP system performance

Solar thermal assisted ASHP system can save energy and benefit environment using sustainable solar energy. [12] At Table 3-2, the main evaluation parameters studies by most researches include COP and energy relative parameters such as energy efficiency, energy consumption and energy saving ratio and so on, i.e., Wu et al. [13] carried out the experiment of lowtemperature ST-ASHP system for space heating. The solar thermal collectors were connected in parallel with the evaporator. The system has a mean COP of 3.2, when it is designed to supply heating to buildings in transition season and the coldest winter days when the ASHP system cannot meet the demand. Only few papers also contain environmental and economic benefits. For example, Kadir et al. [14] calculated the payback time of the solar thermal assisted ASHP according to LPG, electric and fuel oil is 1.4, 2.9 and 3.9 per year, respectively.

These studies are usually conducted under different conditions, and the typical study is Wu et al. [6] and Xu et al. [15-16] Wu et al. [1] set experiment to compare the performance of ASHP and solar assisted ASHP system for domestic hot water (DWH) in different outdoor conditions and then simulated the two systems for the whole year's performance in Shanghai, China. They measured outlet temperature of two different systems in clear and overcast day and night and simulated the month average COP according to the ambient temperature over cloudy/rainy/snowy days. They concluded that the COP of ST-ASHP system performs better especially in the winter, when the ambient temperature is low, since ST-ASHP system draw the heat from the solar can increase the evaporating temperature more remarkably. Xu et al. [15-16] proposed the solar integrated air source heat pump (SIASHP) with R407c. The system is mainly composed of the SIASHP and capillary copper pipe network without any other large component. They completed the experiment in the laboratory and the first generation of prototype of SIASHP was shown in Figure 3-2. In the system, the R407c absorbs the heat from both solar radiation and air source at the same time in the solar finned tube evaporator and runs through the compressor into the capillary copper pipe network, and then goes by the electronic expansion valve back into the solar finned tube evaporator to complete a whole working cycle. The solar finned tube evaporator bypass regulates the internal refrigerant flow of solar finned tube evaporator to avoid the excessive overheating of refrigerant into the compressor protecting the safe operation of compressor. The compressor bypass is the standard configuration for the direct current frequency conversion compressor to protect it from liquid hammer when underload. They claimed the average COP of SIASHP with the part load rate of 100% is 2.77, which is 9.8% higher than 2.52 COP of ASHP. SIASHP has a dramatically better heating performance with a low level of part load rate in the lab, and they also analyzed the energy efficiency of SIASHP system. Taiyuan and a 130 W/m^2 solar irradiance, 76.8% exergy efficiency of SIASHP is 7.9% higher than 71.1% exergy efficiency of ASHP. From 9:00 to 21:00, the 2.89 kWh power consumption of SIASHP is 7.4% lower than 3.12 kWh of ASHP.

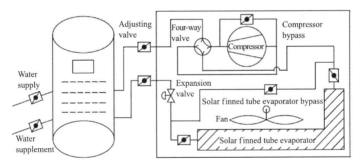

Figure 3-2 Schematic diagram of SIASHP system [28-29]

As for mathematic analysis researches, most of them are under stable condition, which is easy to calculate. However, Cai et al. [17] have built a mathematical model and done the thermodynamic analysis. They tested the performance of IX-SAMHP system by using numerical analysis and compared the experiment results with simulated model in the space heating mode and space water mode condition. Cai [17] and Ji et al. [18] proposed a novel indirect expansion solar-assisted multi-functional heat pump which composes of the multi-functional heat pump and solar thermal collecting system in Hefei, China. The schematic is shown in Figure 3-2. The system can fulfil space heating, space cooling and water heating with high energy efficiency by utilizing solar energy. The refrigerant circulation loop for the solar space heating is 1-2-9-7-6-5-2-3-1 and for the space cooling plus water heating mode is 1-2-5-6-8-9-3-2-1, while the water circulation loops are 12-14-13-12/13-11-5-13 and 5-10-11-5, respectively. Cai et al.[19] investigated experimentally and theoretically on a novel dual source multi-functional heat pump (DMHP) system. The DMHP system can supply air conditioning and domestic water with air source or solar energy in different working modes. In Figure 3-3, the DMHP system contains indoor units and outdoor units linked by refrigerant circuit and water circuit. The refrigerant circuit is a multi-functional heat pump system consisting of two air heat exchangers, a plate-type heat exchanger, a domestic water tank and a compressor. They tested the influence factor which affect the system including the initial solar water tank, the solar irradiation in the solar water heating mode and solar space heating mode. For the space heating mode, the increase of the indoor environment temperature decreases the heating capacity and COP. Increasing the initial water temperature in solar water tank can improve the condensing power and evaporation power, as well as the energy consumption and COP of the system. Moreover, the higher solar irradiation can lead to higher heat

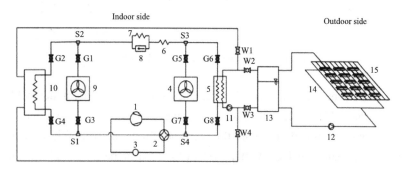

1.compressor; 2.reversing valve; 3.liquid accumulator; 4.outdoor air heat exchanger; 5.plate-type hear exchanger; 6-7.caillary tube; 8.one-way valve; 9.indoor air heat exchanger; 10.domestic water tank; 11-12.water pump; 13.solar water tank; 14.solar flat-palt collector; 15.solar simulator; G1-G8 refrigeration valve; S1-S4 three-way valve; W1-W4 water valve

Figure 3-3   Schematic of the solar-assisted multi-functional heat pump system [19]

transfer rate and larger energy consumption. In annual analysis, DMHP system can obtain relatively high COP of the value above 2.0 throughout the year with the optimal working strategy in three cities under different climate conditions.

One special study among them is that Chen et al. [4] They investigated a solar combi-system consisting of solar collector and a CO_2 heat pump to analyze its performance. Chen et al. [4] conducted the experiment study on the performance of a pre-existing solar combi-CO_2 heat pump system. Figure 3-4 shows a schematic diagram of the system in residential building. The system separates the operation tank from the storage tank. The storage tank acts as the main storage of the captured solar energy and when the temperature level is achieved, it delivers energy to the operation tank. The CO_2 heat pump will consume less electricity to maintain the temperature in the small volume operation tank when solar radiation is poor. The operation tank can also be a buffer of heat for whole system with the obvious temperature changes caused by solar radiant. It can weaken the temperature fluctuation. The result showed that the solar assisted system can save electricity consumption at 1790.8 kWh every year with an average COP of 13.5, comparing with CO_2 HP heating system based on year-round operation (COP is 2.18).

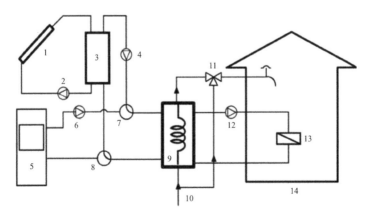

1.solar collector array; 2.solar collector pump; 3.storage tank; 4.solar collector delivery pump; 5.CO_2 heat pump; 6.heat pump delivery pump; 7-8.switching valve; 9.operation tank; 10.civil water inlet; 11.mixing valve; 12.supply pump; 13.fan coil units; 14.testing room

Figure 3-4 Diagram of solar combi-system with CO_2 HP [4]

For extreme condition studies on ST-ASHP system, only a few papers have conducted the related research in recent years. The related research mainly concentrated on the performance of the system by controlling different operation modes or under different ambient conditions. [20] Huang et al. [21-22] experimentally investigated the frosting characteristics of solar collectors with direct expansion air source heat pump system. They conducted the experiment by supporting different temperature, controlling the air humidity and solar irradiation in a lab. As a result, the ambient condition including the temperature, humidity and solar irradiation had a significant effect on the performance of this system under frosting conditions. The frost occurred when the ambient humidity was 50% to 70%, the temperature ranged from 7 °C to 6 °C with zero solar irradiation. While the solar irradiation was 100 W/m^2, there was no frosting except the ambient temperature was lower than −3°C and the humidity is higher than 90%. Long et al. [23] studied the performance of a dual-solar and sir heat source integrated heat pump evaporator in cold season. The hot water of evaporator had an obvious effect on refrigerant's evaporator temperature and COP of the heat pump. Liu et al. [24] found that the heat capacity and COP of solar assisted air heat pump system increased 62% and 59%, compared with air source heat pump system, when the ambient temperature was −15°C.

3.3 Comparison of ST-ASHP systems

In terms of ST-ASHP system, researchers using simulation aim to optimize the system, such as the structure, operation and efficiency, or to verify the performance of system, as some experiments can not achieve in real projects. Weather data mainly containing temperature, solar radiation and seasonal change are the most important input parameters for simulation. The output parameters contain SPF, COP and tank volume and so on.

Most of the simulation studies focus on the optimization performance of ST-ASHP system. Many researches compare their own proposed system with original ASHP system. For example, Jonas et al. [1] developed and validated a tool, SHP-SimFrame in TRNSYS, to simulate the solar heat pump system (SHP), especially solar thermal heat pump. Kim et al. [2] carried out three indirect solar assisted heat pump systems including the serial type solar assisted heat pump, solar assisted heat pump with hybrid solar collectors and parallel solar assisted heat pump system with flat plate solar thermal collectors. Zhang et al. [3] studied the structural parameters on the performance of a direct expansion solar assisted heat pump including the solar collector area, the collector thickness and the pipe length and internal diameter of condenser. And they got the optimized structural parameters for their system.

To make the table clearer, we chose several researches including the media studies except R22 and a few typical studies with new equipment. Almost all the simulation papers through software are conducted in TRNSYS software. i.e., Chen et al. [4] used TRNSYS to optimize ST-ASHP using CO_2 as heat pump refrigerant. The simulated results showed the optimized system has a high performance with the solar fraction can reach 71.1%, except Jonas et al. [1] they developed and validated a tool "SHP-SimFrame" in TRNSYS to simulate the solar heat pump system (SHP), especially solar thermal heat pump. They also only used the simulation method to compare the performance of ST-ASHP system against the conventional ASHP system under various boundary conditions, but the conclusion was not representative, as the simulation parameters changed, the conclusion would also change.

Besides the simulation software method, Li et al. [5] and Deng et al. [6] developed their own mathematic models to simulate performance of ST-ASHP systems, respectively. Both of them calculated the COP value and found that solar irradiation and ambient air temperature have a great effect on the system performance. They indicated that the solar efficiency decreases with the growth of solar irradiation and decreases of ambient temperature. For detail information, Li et al. [7] set up the parameters for components in TRNSYS and calculated SPF and SF of the system. It is proved that the overall efficiency of a DHW system was improved by the combination of a solar collector and heat pump. Deng et al. [6] proposed a modified direct expansion solar assisted air source heat pump water heater system, which is illustrated in Figure 3-5. The proposed system mainly contains a flat-plate solar collector, an evaporator, a compressor, a hot water tank with heat exchanger as a condenser, areceiver and two electronic expansion valves (EEV). There were two operation modes. When the solar radiant is high enough to heat the water efficiently, it run on single solar collector mode. While during the solar radiant is unavailable or low, and evaporating temperature is much lower than ambient air, it works under combination mode. As a result, at solar radiation of 100 W/m^2, the heating time of this system decreases by 19.8% compared to renovation when water temperature reaches 55℃. Meanwhile, the average system COP increases by 14.1%.

3.4 Summary

In this chapter, ST-ASHP system is presented based on existing studies. The system design and main results of these papers are given. Besides that, the system performance common evaluation indicators mainly including COP are compared.

References

[1] Jonas D, Theis D, Felgner F, et al. A TRNSYS-based simulation framework for the analysis of solar thermal and heat pump systems[J]. Applied Solar Energy, 2017, 53: 126-137.

[2] Kim T, Choi B I, Han Y S, et al. A comparative investigation of solar-assisted heat pumps with solar thermal collectors for a hot water supply system[J]. Energy Conversion and Management, 2018, 172: 472-484.

[3] Zhang D, Wu Q B, Li J P, et al. Effects of refrigerant charge and structural parameters on the performance of a direct-expansion solar-assisted heat pump system[J]. Applied Thermal Engineering, 2014, 73(1): 522-528.

[4] Chen J F, Dai Y J, Wang R Z. Experimental and theoretical study on a solar assisted CO_2 heat pump for space heating[J]. Renewable Energy, 2016, 89: 295-304.

[5] Li Y H, Kao W C. Performance analysis and economic assessment of solar thermal and heat pump combisystems for subtropical and tropical region[J]. Solar Energy, 2017, 153: 301-316.

[6] Deng W, Yu J. Simulation analysis on dynamic performance of a combined solar/air dual source heat pump water heater[J]. Energy Conversion and Management, 2016, 120: 378-387.

[7] Li H, Cao C, Feng G, et al. A BIPV/T system design based on simulation and its application in integrated heating system[J]. Procedia Engineering, 2015, 121: 1590-1596.

[8] Fraga C, Hollmuller P, Mermoud F, et al. Solar assisted heat pump system for multifamily buildings: Towards a seasonal performance factor of 5? Numerical sensitivity analysis based on a monitored case study[J]. Solar Energy, 2017, 146: 543-564.

[9] Song M, **a L, Mao N, et al. An experimental study on even frosting performance of an air source heat pump unit with a multi-circuit outdoor coil[J]. Applied Energy, 2016, 164: 36-44.

[10] Song M, Mao N, Xu Y, et al. Challenges in, and the development of, building

energy saving techniques, illustrated with the example of an air source heat pump[J]. Thermal Science and Engineering Progress, 2019, 10: 337-356.

[11] Qiu G, Xu Z, Cai W. A novel integrated heating system of solar energy and air source heat pumps and its optimal working condition range in cold regions[J]. Energy Conversion and Management, 2018, 174: 922-931.

[12] Nemati O, Ibarra L M C, Fung A S. Review of computer models of air-based, curtainwall-integrated PV/T collectors[J]. Renewable and Sustainable Energy Reviews, 2016, 63: 102-117.

[13] Wu J, Chen C, Pan S, et al. Experimental study of the performance of air source heat pump systems assisted by low-temperature solar-heated water[J]. Advances in Mechanical Engineering, 2013, 5: 843013.

[14] Bakirci K, Yuksel B. Experimental thermal performance of a solar source heat-pump system for residential heating in cold climate region[J]. Applied Thermal Engineering, 2011, 31(8-9): 1508-1518.

[15] Dong X, Tian Q, Li Z. Energy and exergy analysis of solar integrated air source heat pump for radiant floor heating without water[J]. Energy and Buildings, 2017, 142: 128-138.

[16] Dong X, Tian Q, Li Z. Experimental investigation on heating performance of solar integrated air source heat pump[J]. Applied Thermal Engineering, 2017, 123: 1013-1020.

[17] Cai J, Ji J, Wang Y, et al. Numerical simulation and experimental validation of indirect expansion solar-assisted multi-functional heat pump[J]. Renewable Energy, 2016, 93: 280-290.

[18] Jie J, **gyong C, Wenzhu H, et al. Experimental study on the performance of solar-assisted multi-functional heat pump based on enthalpy difference lab with solar simulator[J]. Renewable Energy, 2015, 75: 381-388.

[19] Cai J, Ji J, Wang Y, et al. Operation characteristics of a novel dual source multi-functional heat pump system under various working modes[J]. Applied Energy, 2017, 194: 236-246.

[20] Mohamed E, Riffat S, Omer S. Low-temperature solar-plate-assisted heat pump: A developed design for domestic applications in cold climate[J]. International Journal of Refrigeration, 2017, 81: 134-150.

[21] Huang W, Ji J, Xu N, et al. Frosting characteristics and heating performance of

a direct-expansion solar-assisted heat pump for space heating under frosting conditions[J]. Applied Energy, 2016, 171: 656-666.

[22] Huang W, Zhang T, Ji J, et al. Numerical study and experimental validation of a direct-expansion solar-assisted heat pump for space heating under frosting conditions[J]. Energy and Buildings, 2019, 185: 224-238.

[23] Long J, Zhang R, Lu J, et al. Heat transfer performance of an integrated solar-air source heat pump evaporator[J]. Energy Conversion and Management, 2019, 184: 626-635.

[24] Liu Y, Ma J, Zhou G, et al. Performance of a solar air composite heat source heat pump system[J]. Renewable Energy, 2016, 87: 1053-1058.

[25] Poppi S, Bales C. Techno-Economic Analysis of a Novel Solar Thermal and Air-Source Heat Pump System[J]. 2016.

[26] Poppi S, Bales C, Haller M Y, et al. Influence of boundary conditions and component size on electricity demand in solar thermal and heat pump combisystems[J]. Applied Energy, 2016, 162: 1062-1073.

[27] Fraga C, Mermoud F, Hollmuller P, et al. Large solar driven heat pump system for a multifamily building: Long term in-situ monitoring[J]. Solar Energy, 2015, 114: 427-439.

[28] Ni L, Qv D, Shang R, et al. Experimental study on performance of a solar-air source heat pump system in severe external conditions and switchover of different functions[J]. Sustainable Energy Technologies and Assessments, 2016, 16: 162-173.

[29] Niu F, Ni L, Yao Y, et al. Performance and thermal charging/discharging features of a phase change material assisted heat pump system in heating mode[J]. Applied Thermal Engineering, 2013, 58(1-2): 536-541.

Chapter 4
Research of PV-ASHP System

4.1 Research orientation

For PV-ASHP, it is similar that most authors compare the energetic performance of their proposed system with original system or other systems. For example, Evangelos et al.[1] compared ASHP with PV modules and WSHP with flat plate collectors (FPC) and PVT system. The relevant system simulation methods and results are summarized in Table 4-1. They compared between ASHP with PV and PV/T for space heating. Giuseppe et al[2] set four simulations to compare the energy performance of ASHP with PV modules system and conventional ASHP system by using TRNSYS and Matlab. They assessed the performance in a detached house located in northern Italy with two different cases: one with a PV module and the conventional tile without PV. Their work was the application of a title for under-slating ventilation ducts, which was modeled in TRNSYS. They found that solar and air source combined heat pump system demonstrates as the most advantageous case compared to conventional ASHP systems.

While for economic performance, the related studies is few. Evangelos et al.[1] examined an ASHP system with PV modules for space heating proposes. Figure 4-1 depicts the system concept. The electricity demand of the heat pump is covered partially by PV panels and partially by grid electricity import. To convert and regulate the voltage and store the momentary supplementary energy, an inverter and batteries are used in this system. They developed the related TRNSYS models and compared the performance, the electricity consumption and financial performance among four systems, which are PV-ASHP systems, water source heat pump with flat plate collectors (FPC-WSHP), water source heat pump with thermal photovoltaic collectors (PV/T-WSHP) and water source heat pump with photovoltaic and flat plate collectors (PV+FPC-WSHP). They concluded that when the electricity cost is between 0.2 €/kWh and 0.23 €/kWh, the use of 20 m² PV area with an air source heat pump is the most attractive solution financially in Greece.

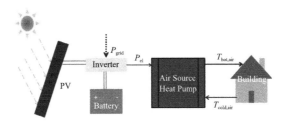

Figure 4-1 Air source heat pump heating system coupled with PV modules [3]

4.2 PV-ASHP system performance

Summary of experiment methods and the related results of PV-ASHP systems in Table 4-2, the common researches for PV-ASHP system also concentrate on the energetic performance, i.e., Giampaolo et al.[4] used the experiments to assess the performance of a PV-ASHP system and validate their simulation model.They [5] adopted a methodology similar to that described in to measure the flow rate and temperature of the air at the outlet of the ventilation channel, integrating these measurements into a comprehensive analysis. They found that the average COP increased from 3.60 to 3.75, turning into a primary energy reduction with respect to the reference case of 5%.

Besides the energetic performance, some studies also evaluated the environmental and economic performance. Wang et al. [6] set up an experiment of the integrated PV-ASHP system and tested the system in six cases with different PV powers. Their integrated PV-ASHP system was made up of three sub-systems: a PV system, an electricity storage, an inverter system and an ASHP system, and the experiment located on the top floor of a building. They indicated that PV-ASHP system could have a saving rate of 41.16% of exergy consumption per unit investment for cooling and 35.02% for heating. The life expectancy of PV-ASHP system could reach about 26

Table 4-1 Summary of simulation methods and the related results of PV-ASHP systems

Reference	System	Tool	Model	Location	Operation conditions	Input data	Building types	Evaluation parameters	Results
[1]	PV-ASHP; PVT-WSHP	TRNSYS	Type 56-building; Type 655-3-heat pump	Greece	heating mode	undefined parameter; the collecting area and storage tank volume	commercial building	monthly load; COP; energy consumption	▪lower energy consumption than ASHP and fan coil heating system; ▪Indoor temperature is better in solar driven heat pump system; ▪The COP is near 4 and for the conventional air source systems close to 2.5

Table 4-2 Summary of experiment methods and the related results of PV-ASHP systems

Reference	System	Medium	Operation mode	Location	Measured parameters	Building types	Evaluation parameters	Main results
[6]	PV-ASHP	—	heating/cooling mode	Changsha, Central South China	different PV power	office building in university	exergy efficiency/ consumption; life expectancy; CO_2 emission reduction	▪life expectancy of PV-ASHP system in Central South China is about 26 years; ▪Install PV capacity decided by ASHP required rating power; ▪PV-ASHP has an energy consumption saving rate of 41.16% for cooling and 35.02% for heating

Chapter 4
Research of PV-ASHP System

Continued

Reference	System	Medium	Operation mode	Location	Measured parameters	Building types	Evaluation parameters	Main results
[7]	PV-AC	R134a	heating mode	Alicante, Spain	irradiation; environment temperature; air mass flow rate	residential buildings	heat recovered; heat production; COP; energy performance index; primary energy consumption	• Average COP of heat pump increases from 3.6 to 3.75; • PV cell integrated in the tile can produce more energy than air conditioning required
[4]	PV module traditional HP system	—	Real operation condition VS. a scaled-down channel (1:5)	Northern Italy	the integral absorptivity and emissivity; the airflow pressure drop	residential building	Air temperature; Energy consumption; Integration angle; COP	• air temperature variation ranges from 2 °C to 20 °C leading to a heat recovered between 2 kW and 7 kW in winter and summer time, respectively; • the integration of the PV cell in the tile reduces the thermal energy by 15%; • The heat recovered enhances the heat pump COP, reducing the Primary Energy Consumption of a conventional heat pump by 5%, and achieves with limited additional costs

years and it can reduce the 11.10 t of CO_2 in the whole life emission and save the operation cost comparing with ASHP powered by electricity from the national grid.

Besides the conventional PV/T-ASHP system, Aguilar et al. [7] conducted an experiment using photovoltaic air conditioning unit (PV+AC). They carried out the experiment on an air conditioning unit which had been powered using both a PV installation and the grid simultaneously for a whole year with the control system, which was designed to maximize solar contribution in Spain. The PV+AC system consisted of the air conditioning unit with two electrical connections (PV panels and grid) and the PV installation. They tested PV+AC system from May to October in cooling mode in an office located in Alicante (Spain) and the whole system had demonstrated to be 100% reliable, having undergone no maintenance. Then they optimized the operation mode, regulating the air conditioning unit independently its own operation regime to maximize the PV energy input. The solar contribution obtained in cooling mode from May to October was 64.5%, while the production factor was 65.1%. The performance of the system will depend on the solar radiation, outlet temperature and the unit's load factor.

4.3 Summary

In this chapter, PV-ASHP system is presented based on existing studies. The system design and main results of these papers are given. Besides that, the system performance common evaluation indicators mainly including COP are compared.

References

[1] Bellos E, Tzivanidis C, Moschos K, et al. Energetic and financial evaluation of solar assisted heat pump space heating systems[J]. Energy conversion and Management, 2016, 120: 306-319.

[2] Manzolini G, Colombo L P M, Romare S, et al. Tiles as solar air heater to support a heat pump for residential air conditioning[J]. Applied Thermal Engineering, 2016, 102: 1412-1421.

[3] Kamel R S, Fung A S. Modeling, simulation and feasibility analysis of residential BIPV/T+ ASHP system in cold climate—Canada[J]. Energy and Buildings, 2014, 82: 758-770.

[4] Manzolini G, Colombo L P M, Romare S, et al. Tiles as solar air heater to support a heat pump for residential air conditioning[J]. Applied Thermal Engineering, 2016, 102: 1412-1421.

[5] Dubey S, Sandhu G S, Tiwari G N. Analytical expression for electrical efficiency of PV/T hybrid air collector[J]. Applied Energy, 2009, 86(5): 697-705.

[6] Wang C, Gong G, Su H, et al. Efficacy of integrated photovoltaics-air source heat pump systems for application in Central-south China[J]. Renewable and Sustainable Energy Reviews, 2015, 49: 1190-1197.

[7] Aguilar F J, Aledo S, Quiles P V. Experimental analysis of an air conditioner powered by photovoltaic energy and supported by the grid[J]. Applied Thermal Engineering, 2017, 123: 486-497.

Chapter 5
Research of PV/T-ASHP System

5.1 Research orientation

Similar with all ASHP multi-system, the simulation tool for PV/T-ASHP system is also TRNSYS. And the performance of PV/T-ASHP system are the main studies' contents, the main methods are to change the parameters including the ambient condition or operation modes. They also proposed that, there exist technical problems to hinder PV/T collector to be used in large scale. For example, Raghad et al [1-2] proved in their TRNSYS model that an integrated BIPV/T+ASHP thermal energy system increases the overall COP and improves the energy efficiency of buildings.

Besides TRNSYS simulation, Cao et al. [3] used the CFD simulation to test the performance of PV/T-ASHP, which mainly studies the outlet temperature of the collector and the COP of ASHP operated in severe cold region. Simulation time step was set to be 0.25h, tolerance integration and convergence were both 0.01. The control scheme was based on the load profile for the system. They obtained the result for supporting heating that the outlet temperature of PV/T collector couid reach 76.6℃ and the COP of ASHP unit reached 4.1, which increased even in the low outdoor temperature.

Different from other studies that used simulation software, Getu et al. [4] did the theoretical investigation of the performance of a two-stage variable capacity air source heat pump coupled with a BIPV/T system. They used the simulation method, and the BIPV/T system model was a TRNSYS type 568 component. The main results they found that the results were different along with ambient temperature. The COP of the new system increased obviously when the average temperature was between −3℃ to 10℃ ; otherwise, there was no significant change.Table 5-1 presents a summary of the simulation methods and their associated results for PV/T-ASHP systems. Similarly,

Chapter 5
Research of PV/T-ASHP System

Table 5-1 Summary of simulation methods and the related results of PV/T-ASHP systems

Reference	System	Tool	Model	Location	Operation conditions	Input data	Building types	Evaluation parameters	Results
[3]	PV-LHP/ SAHP	mathematical model	—	Qinhuangdao, China	solar/air source heat pump	solar irradiation and ambient air temperature	residential buildings	operation performance of under typical working conditions; economic feasibility analyses	• solar energy utilization efficiency decreases with solar irradiation growth and ambient temperature decrease; • monthly average COP of heat pump modesis 3.10; • The annual solar heating ratio of the system is up to 57.8%; • Life cycle cost of the PV-LHP/SAHP could be reduced by 29.6% than ASHP
[4]	ASHP VS. ASHP BIPV/T	TRNSYS	Type 568-BIPV/T system; Type 56-multi-zone building; Type 687-a window model	Anchorage, Southern Alaska, USA	the ambient air supplied to HP; the warm air coming out of BIPV/T is supplied into HP	irradiation	residential building	COP	• COP improved for average ambient temperature is above $-3\ ℃$; • COP can not be improved, when average ambient temperatures are above 10 ℃ or below $-10\ ℃$; • The maximum COP of BIPV-ASHP is 5.31

ASHP Assisted Solar Energy Systems:
Classification, Applications, and Performance Research

Continued

Reference	System	Tool	Model	Location	Operation conditions	Input data	Building types	Evaluation parameters	Results
[1]	BIPV/T-ASHP	TRNSYS	—	Toronto, Canada	cooling/heating/DHW/floor heating	air flow rate	test hut	COP	• An integrated BIPV/T+ASHP thermal energy system increases the overall performance; • the seasonal COP could be increased from 2.74 to 3.45. • The heat pump electricity consumption is reduced by 20% for winter
[2]	BIPV/T-ASHP	TRNSYS 17	Type 567-BIPV/T; Type 56-building model; Type 665-ASHP	Ontario, Canada	heating mode	outdoor temperature	Archetype Sustainable House	COP; saving in energy and cost; GHG emission reduction	• The heat pump electricity consumption is reduced by 20% for winter; • GHG emission reduced and energy consumption got saving; • COP increased from 2.74 to 3.45
[3, 9]	PV/T-ASHP	CFD	Gambit; Fluent	Shenyang, China	steady-state RNG k-ε turbulence model	inlet and outlet velocity	residential building	efficiency of BIPV/T; COP	• COP of heat pump unit reaches 4.6; • Thermal efficiency of BIPV/T-ASHP integrated heating system is relatively high in low temperature environment

Chapter 5
Research of PV/T-ASHP System

Table 5-2 Summary of experiment methods and the related results of PV/T-ASHP systems

Reference	System	Medium	Operation mode	Location	Measured parameters	Building types	Evaluation parameters	Main results
[3]	PV/T-ASHP	R410A	cooling mode	Shenyang, China	weather, thermal load, solar irradiance	residential building	power consumption; energy consumption; solar contribution; production factor; load factor; useful thermal energy; EER	• System performance depended on the environment including solar radiation, outlet temperature and the unit's load factor; • Several technological improvements in the analyzed system have been made to optimize the system performance
[2]	PV/T-ASHP with TES	—	energy efficiency / energy consumption cost model	Toronto, Canada	—	Test hut	model relative error	• The relative error of exergy efficiency mode of integrated PV-ASHP for cooling is less than 4.21%; • The relative error of exergy consumption cost model is 1.5% for cooling and 0.3% for heating
[10]	PV/T-ASHP	—	hot water	Beijing, China	ambient temperature and solar irradiance	laboratory	COP	• COP of heat pump reduces from 5.61 to 1.69 and the average is 3.03. • Comprehensive system COP ranges from 6.07 to 1.33, the average is 2.99

Table 5-2 outlines the experimental methods used and the corresponding results for these systems.

5.2 Experiment setting of PV/T-ASHP system

To improve the heat transfer performance and efficiency, in this chapter, a novel construction of the heat exchanger including an ultra-thin superconducting thermal absorber with internal fins shaped bubbling with two different sizes is proposed. This solar thermal absorber has two different sizes of fins, in contrast with the existing similar solar thermal absorber [5-6], this novel structure solar thermal absorber has lower resistance and the turbulence for the liquid in the absorber is higher. This is because of its design and fins' structure. The width for this absorber is only 4 mm, which is much smaller than the original absorbers. The capacity is also reduced, so the resistance becomes lower. Besides, the fins increase the liquid turbulence compared with the original flat absorber type. [5] The novel absorber is also studied using simulations and the novel absorber model is built for CFD. The performance of the novel absorber is further simulated for different boundary conditions.

Based on the research conducted by our group [5], the performance of this kind of absorber is better than the original types. However, only one structure is considered in that work. [5] In this chapter, seven different fin shapes models of the thermal absorbers are also built to make a comparison with original absorbers. The performance of these thermal absorbers including the temperature and the efficiency is further analyzed. The absorber with the best performance is then chosen to be further investigated using simulation and experiment methods. This research aims to propose a novel thermal absorber and evaluate its performance for future applications.

Firstly, the construction, the corresponding analytical method, and simulation boundary conditions in the CFD model are introduced. Then, the simulation results for different conditions and different constructions are presented and discussed.

5.2.1 Simulation models of the solar absorber

To compare the impact of the different shapes on thermal performance, different construction absorbers models are built. We consider the following seven models, the original type with big and small bubbling, the flat absorber, deleting small bubbling absorber based on the original type, deleting the big bubbling and leaving small absorber based on the original type, changing all the bubbling to big, changing all bubbling into small, and changing the whole width of the absorber from 4 mm to 5 mm, total seven types. For a fair comparison, the design sizes are kept constant and equal to the original type as shown in Figure 5-1. The boundary condition is also different from the original three modes. The boundary condition of the absorber surface connected to the PV collector is constant, 30°C, the heat flux for other surfaces is set to 0, changing the inlet water velocity from 0.2 m/s to 1 m/s, per 0.1 m/s variations. The outlet boundary condition is the outflow.

In the CFD model building process, a realizable k-ε turbulence model is chosen to calculate the thermal performance of the novel absorber. The realizable k-ε turbulence model originates from the statistical data, similar to the standard k-ε turbulence model. However, the realizable k-ε model differs from the standard k-ε model in the following two aspects. First, it contains a new formulation for the turbulent viscosity C_μ. Furthermore, in contrast to the standard model, C_μ is variable. The second difference is a new transport equation for the dissipation rate, ε, which is derived from an exact equation for the transport of the mean-square vorticity fluctuation. As a result, the realizable k-ε turbulence model mainly provides improved predictions for the spreading rate of jets and has a superior ability to capture the mean flow of complex structures and for flows involving rotation, boundary layers under strong adverse pressure gradients, separation, and re-circulation. The equations for the realizable k-ε turbulence model are [6-7]:

ASHP Assisted Solar Energy Systems:
Classification, Applications, and Performance Research

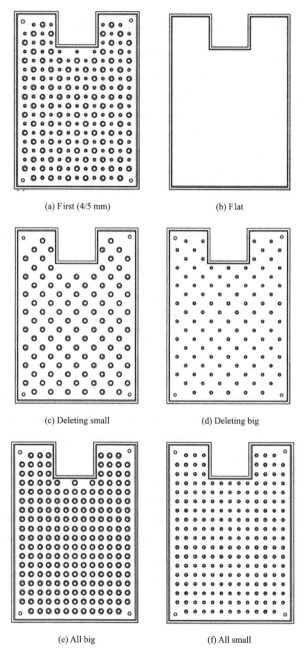

(a) First (4/5 mm) (b) Flat
(c) Deleting small (d) Deleting big
(e) All big (f) All small

Figure 5-1　Different fin shapes of the thermal absorber

$$\frac{\partial}{\partial t}(\rho k) + \frac{\partial}{\partial x_i}(\rho k u_i) = \frac{\partial}{\partial x_i}\left[\left(\mu + \frac{\mu_t}{\sigma_k}\right)\frac{\partial k}{\partial x_i}\right] + G_k + G_b - \rho\epsilon - Y_M + S_k \quad (5\text{-}1)$$

$$\frac{\partial}{\partial t}(\rho\epsilon) + \frac{\partial}{\partial x_j}(\rho\epsilon u_j) \quad (5\text{-}2)$$

$$= \frac{\partial}{\partial x_j}\left[\left(\mu + \frac{\mu_t}{\sigma_\epsilon}\right)\frac{\partial\epsilon}{\partial x_j}\right] + \rho C_1 S_\epsilon - \rho C_2 \frac{\epsilon^2}{k + \sqrt{\nu\epsilon}} + C_{1\epsilon}\frac{\epsilon}{k} C_{3\epsilon} G_b + S_\epsilon$$

$$C_1 = \max\left[0.43, \frac{\eta}{\eta + 5}\right] \quad (5\text{-}3)$$

$$\eta = S\frac{k}{\epsilon} \quad (5\text{-}4)$$

$$S = \sqrt{2S_{ij} S_{ij}} \quad (5\text{-}5)$$

$$\mu_t = \rho C_\mu \frac{k^2}{\epsilon} \quad (5\text{-}6)$$

where, u_i represents the velocity component in the corresponding direction; μ_t denotes eddy viscosity; G_k represents the generation of turbulence kinetic energy due to the mean velocity gradients, calculated in the same manner as the standard k-ε turbulence model; G_b is the generation of turbulence kinetic energy due to buoyancy; $C_{1\epsilon}$=1.44, and C_2=1.9.

In this study, the realizable k-ε turbulence model in Fluent is selected to build the thermal collector model in standard wall functions with water as the liquid materials. The simulation tool is Fluent 6.3.

Through simulation, the thermal performance of these different types of absorbers is analyzed. The heat collecting and thermal efficiency of these absorbers are calculated by the equations as presented in equation (5-7) and equation (5-8)

$$Q_C = m_w \times (T_{out} - T_{in}) \quad (5\text{-}7)$$

where, m_w is the water mass volume, T_{out} is the outlet water temperature and T_{in} is the inlet water temperature of this absorber.

The results of the heat collecting and the efficiency of each type are shown in Figure 5-2 and Figure 5-3. For different shapes of the absorbers, the trend of the heat collecting and the efficiency η is the same as the variation of the inlet water velocity. The higher the inlet water velocity, the higher the heat collecting, and the lower the efficiency η. Taking the first absorber type as an example, by increasing the inlet water velocity from 0.2 m/s to 1 m/s, the heat collecting is changed from 15×10^6 J/h to 49.5×10^6 J/h, however, the efficiency η, is reduced from 66% to 25%. The performance of the absorbers with bubbling is much higher than that of the flat type both in terms of heat collecting and efficiency aspects.

For heating collection, increasing the inlet water velocity increases the heating collecting for all seven types of absorbers. For the first absorber type, the values vary from 15×10^6 J/h to 49.5×10^6 J/h. By deleting the small type the values are changed from 15.2×10^6 J/h to 44.5×10^6 J/h, and by deleting the big type the values are increased from 15.1×10^6 J/h to 49.2×10^6 J/h. All big type gains from 14.8×10^6 J/h to 48.4×10^6 J/h, all small type increases from 14.7×10^6 J/h to 49.5×10^6 J/h. Changing the width to 5 mm type, the heat collecting increases from 16×10^6 J/h to 40.8×10^6 J/h, while the flat type is the lowest, 9×10^6 J/h to 36×10^6 J/h.

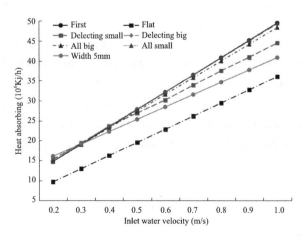

Figure 5-2 Heat absorbing per hour of different construction collectors (10^6 kJ/h)

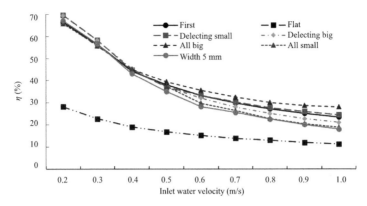

Figure 5-3 The efficiency of different construction absorbers (%) versus inlet water velocity

In terms of efficiency, the efficiency of the heat transfer is decreased by increasing the inlet water velocity and its rate of reduction is decreased by the velocity. When the velocity is larger than 0.6 m/s, the rate is much lower than before. The efficiency value for the first type is also reduced from 66% to 25%, in the deleting small type, the values are reduced from 69.6% to 24.5% and for the deleting big type, it is decreased from 69.5% to 21%. While for all big and all small types, the values are reduced from 96.6% to 58% and from 65.8% to 18.8%, respectively. Changing the width to 5 mm, the efficiency varies from 67% to 18%, and the lowest is the flat type, where it goes from 28% to 11%.

Note that the gradient is reduced by increasing the velocity. By increasing the velocity to larger than 0.6 m/s, the rate becomes much slower than before. When the inlet water velocity is 0.5 m/s, which is suitable considering both heat collecting and efficiency of all these seven types of absorbers, the first type has the second-highest heat collecting capacity (27.8×10^6 J/h) and efficiency (68%) amongst all seven types, while the efficiency for other types is 16.8%, 37%, 39%, 37.3%, 34.9%, respectively, except for the all big type which is 39%, but the difference is rather small. Moreover, considering that the all big type needs higher power in real application with bigger resistance than the first type, the first novel thermal absorber is a better design compared with the other shapes.

5.2.2 Simulation models of the designed absorber

1. The boundary conditions of the simulation

A series of boundary conditions are set to simulate the thermal performance of the ultra-thin superconducting absorber, see Table 5-3. The boundary conditions presented in Table 5-3 include the variations in several aspects such as (1) the heat flux (I) in the ranges from 400 W/m^2 to 1,000 W/m^2; (2) the inlet water temperature ranges from 10°C to 35°C; (3) the velocity of inlet water ranges from 0.2 ms to 1 m/s. The base set is under 800 W/m^2, the inlet water temperature is 20 °C, and the inlet water velocity is set to 0.5 m/s. [8]

The materials of the absorber are steel and aluminum, the value of thermal conductivity for these materials is 16.27 W/m°C, and 237 W/m°C, respectively. The realizable k-ε turbulence model is used with a turbulent intensity of 5% and the viscosity ratio is 10. Density and other thermal properties are assumed to be constant. All models are meshed by using 1mm triangles, while for comparison, 2 mm and 3 mm triangles are also meshed to verify the accuracy of the simulation.

Table 5-3 List of simulation models for the novel ultra-thin superconducting absorber [8]

Simulation mode	I (W/m^2)	T_{in} (°C)	V_{in} (m/s)
1	400 500 600 700 800 900 1,000	20	0.5
2	800	10 15 20 25 30 35	0.5

Continued

Simulation mode	I (W/m^2)	T_{in} (°C)	V_{in} (m/s)
3	800	20	0.2 0.3 0.4 0.5 0.6 0.7 0.8 0.9 1

2. Simulation results

Using CFD simulation, the impact of operational parameters (heat flux, inlet water temperature, and inlet water velocity) on the novel ultra-thin superconducting solar absorber is investigated. We further evaluate the thermal performance of different shapes with the same materials and compare them for different inlet water velocities. Furthermore, we compare the performance based on the outlet water temperature and the map of the whole plate temperature, the outlet water temperature, and the difference between inlet and outlet water temperature are calculated. In this study, the heat loss to the surrounding through radiation transfer is ignored for brevity, and the efficiency of the thermal absorber is calculated by equation (5-8).

$$\eta_{th} = \frac{Q_u}{IAt} = \frac{cm\Delta T}{IAt} = \frac{cm(T_{out}-T_{in})}{IAt} \tag{5-8}$$

We consider three different mesh sizes because it can cover the errors caused by using only one size mesh. Another reason for using different mesh sizes is to compare the result and select the best mesh for the model. As the width of the novel solar thermal absorber is 4 mm, using 1 mm mesh, the result is more accurate and calculation time is reasonable. The simulation convergence results for all three mesh sizes are accurate and for 1 mm mesh size, the convergence result is shown in Figure 5-4.

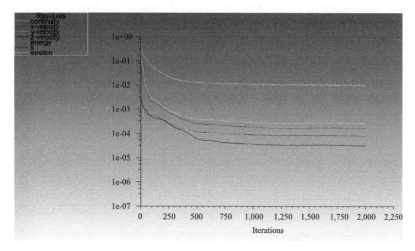

Figure 5-4 Simulation result with 1 mm mesh size

3. Impact of the heat flux

The simulation results are shown in Figure 5-6, where the heat flux is changed from 400 W/m² to 1,000 W/m², with 100 W/m² increase steps while other parameters are kept constant, i.e., inlet water temperature is 20°C and its velocity is 0.5 m/s. The results about the outlet water temperature, the inlet, and outlet water difference, and the efficiency of the absorber are also shown Figure 5-5, and the temperature and pressure distribution examples under solar irradiation with 800 W/m² are shown in Figure 5-6. It is seen that by increasing the heat flux, the outlet water temperature is increased while the inlet water temperature is constant. It is also seen that the temperature difference (ΔT) between the inlet and outlet water is increased from 0.034°C to 0.41°C. Higher heat flux enhances the solar heat transfer of the absorber, causing a significant improvement of the efficiency η from 10.8% to near 60%. As a result, the heat flux has a significant impact on the efficiency η of the ultra-thin superconducting absorber. But considering the lower outlet water temperature which is better for the PV collector, the heat flux should be not so high in real applications. Note that the heat flux is directly related to the solar irradiation.

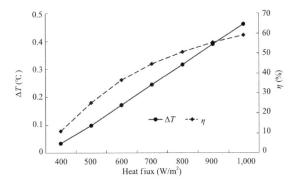

Figure 5-5 Thermal performance of water temperature difference between the inlet and outlet (ΔT), the efficiency (η) for novel absorber in mode 1

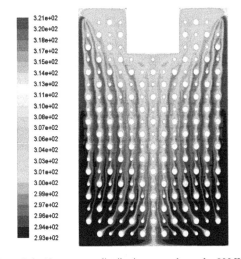

Figure 5-6 Temperature distribution examples under 800 W/m²

4. Impact of inlet water temperature

For simulation mode 2 the simulation results are presented in Figure 5-6, where the inlet water temperature is varied from 10°C to 35°C with the increased steps of 5°C, and heat flux 800 W/m² and inlet water velocity 0.5 m/s. For mode 2, the temperature and the map of the velocity are simulated and the outlet water temperature is recorded in the simulation result. The thermal performance results are shown in Figure 5-7.

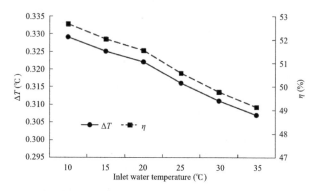

Figure 5-7 Thermal performance of the water temperature difference between inlet and outlet (ΔT), the efficiency for novel absorber under mode 2, η

The outlet water temperature is increased by increasing the inlet water temperature. Different from the impact of the heat flux on the thermal performance of the efficiency, by increasing the inlet water temperature, the difference between the inlet and the outlet temperature is decreased. This results in decreasing the efficiency of the ultra-thin superconducting absorber from 52.7 % to 49 %. The trend is opposite with the heat flux impact and the degree is also different. The impact of the inlet water temperature also affects the efficiency (η) of the absorber. Our results also suggest a low impact of the inlet temperature on the performance of the absorber. Nevertheless, improving the temperature of the inlet water is often costly. Therefore, for the real applications, from the absorber's point of view, the natural water could be directly used as the inlet water.

5. Impact of the inlet water velocity

For simulation mode 3 which is shown in Figure 6-2, the inlet water velocity is changed from 0.2 m/s to 1.0 m/s, with 0.1 m/s steps. Similar to other simulation modes we also keep other parameters constant, i.e., heat flux is 800 W/m² and inlet water temperature is 20°C. In this simulation model, the temperature and the map of the velocity are simulated and the outlet water temperature is recorded in the simulation results. The thermal performance results are shown in Figure 5-8.

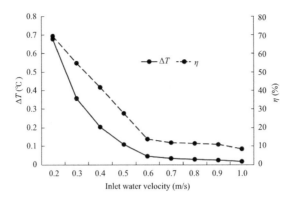

Figure 5-8 Thermal performance of the novel absorber for mode 3

The inlet water velocity highly affects the efficiency of the novel absorber. With the increase of the inlet water velocity from 0.2 m/s to 1 m/s, the efficiency is reduced from 70% to under 10%. For the inlet water velocity larger than 0.6 m/s, the temperature difference between the inlet water and outlet water is close to 0. This means the water in the absorber does not gain enough heat before it is pushed out through the outlet. Moreover, for higher inlet water velocity, the heat transfer distribution is increased. Therefore, the heat transfer resistance is also increased; hence the thermal efficiency is decreased. However, if the water velocity is too low, although the efficiency is high enough, the heat transfer capacity is low. In practice, the inlet water velocity 0.5 m/s might be suitable considering the efficiency and the using time for the collector.

References

[1] Kamel R, Ekrami N, Dash P, et al. BIPV/T+ ASHP: technologies for NZEBs[J]. Energy Procedia, 2015, 78: 424-429.

[2] Kamel R S, Fung A S. Modeling, simulation and feasibility analysis of residential BIPV/T+ ASHP system in cold climate—Canada[J]. Energy and Buildings, 2014, 82: 758-770.

[3] Cao C, Li H, Feng G, et al. Research on PV/T–air source heat pump integrated heating system in severe cold region[J]. Procedia Engineering, 2016, 146: 410-414.

[4] Hailu G, Dash P, Fung A S. Performance evaluation of an air source heat pump coupled with a building-integrated photovoltaic/thermal (BIPV/T) system under cold climatic conditions[J]. Energy Procedia, 2015, 78: 1913-1918.

[5] Shen J, Zhang X, Yang T, et al. Characteristic study of a novel compact Solar Thermal Facade (STF) with internally extruded pin–fin flow channel for building integration[J]. Applied Energy, 2016, 168: 48-64.

[6] Wikipedia. 2023. K-epsilon turbulence model[EB/OL]. Available: https://en.wikipedia.org/wiki/K-epsilon_turbulence_model [2024-04-17].

[7] Aldali Y, Muneer T, Henderson D. Solar absorber tube analysis: thermal simulation using CFD[J]. International Journal of Low-Carbon Technologies, 2013, 8(1): 14-19.

[8] Xu P, Zhang X, Shen J, et al. Parallel experimental study of a novel super-thin thermal absorber based photovoltaic/thermal (PV/T) system against conventional photovoltaic (PV) system[J]. Energy Reports, 2015, 1: 30-35.

[9] Li H, Cao C, Feng G, et al. A BIPV/T system design based on simulation and its application in integrated heating system[J]. Procedia Engineering, 2015, 121: 1590-1596.

[10] Wang G, Quan Z, Zhao Y, et al. Experimental study of a novel PV/T-air composite heat pump hot water system[J]. Energy Procedia, 2015, 70: 537-543.

Chapter 6
Experiment Study of PV/T-ASHP System in an Office Room Application

6.1 Design of PV/T-ASHP system

6.1.1 The detailed information about an office room

The experiment considers an office room which is located in Beijing and its size is shown in Figure 6-1. The length is 12.5 m, and the width is 3 m and the height is 2.9 m. This is an office room with four workers which is used for daily work. The main body of the assembly room is a 10cm polystyrene board with a heat transfer coefficient of 0.04 W/($m^2 \cdot K$). The room has a broken bridge aluminum door with a 330mm × 2,510 mm window and a heat transfer coefficient of 1.75 W/($m^2 \cdot K$). The size of the sliding door is 1,120 mm × 1,100 mm which includes a plastic steel window, hollow glass with anti-theft steel mesh. The size of the steel door is 845mm × 2,025 mm. Ignoring the influence of the column, according to the GB 50736—2012 *Code for Design of Heating, Ventilation and Air Conditioning of Civil Buildings*, the indoor design parameter is 18℃, and the outdoor design parameter is −7.5 ℃.

The building has four walls (east, south, west, and north). The height of the ceiling is 2.9 m. Therefore

The area of the east outer wall is: 2.99×2.896=8.66 m^2

The area of the west outer wall is: 2.99×2.896=8.66 m^2

The area of the south outer wall is: 12.12×2.896 − 0.845×2.025×2 − 1.12×1.1×2=29.2 m^2, the south outer window: 1.12×1.1×2=2.46 m^2, the door: 0.845×2.025×2=3.4 m^2

The area of the north outer wall is: 12.12 × 2.896=35.1 m^2

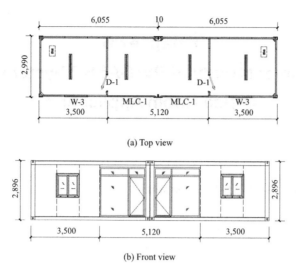

Figure 6-1　Size of the experiment office room (units: m)

Roof: 2.99×12.12=36.2 m²

1. The basic heat consumption of the enclosure structure is:

$$q = KF \times (t_n - t_w) \partial$$

where, t_n is the design temperature for this office and t_w is the design ambient temperature. K is the heat transfer coefficient and F is the area.

The basic heat consumption of the external wall:
$Q_1 = 0.04 \times (0.95 \times 8.66 \times 2 + 0.8 \times 29.2 + 0.95 \times 35.1) \times (18+7.5) = 74.6$ W

Basic heat consumption of exterior windows and doors:
$Q_2 = 1.75 \times 2.46 \times 0.8 \times (18+7.5) = 87.8$ W
$Q_3 = 1.75 \times 3.4 \times 0.8 \times (18+7.5) = 121.4$ W
Roof: $Q_4 = 0.04 \times 36.2 \times (18+7.5) = 36.9$ W

2. Cold air penetration: according to the gap method, the effect of the heat pressure and outdoor wind speed with height is not considered

The building has only south-facing gaps, including:
Window: (1.12×4+1.1×4+1.1×3)×2=24.36 m
Door: (0.845×4+2.025×7)×2=35.11 m

Air infiltration volume:

$V = \sum(lLn) = 24.36 \times 0.5 \times 0.15 + 35.11 \times 0.5 \times 2 \times 0.15 = 7.1 \text{m}^3/(\text{h} \cdot \text{m})$

$Q_5 = 0.278 \times V\rho \times (t_n - t_w) c$

$Q_5 = 0.278 \times 7.1 \times (18+7.5) \times 353/(375 - 7.5) = 48$ W

3. Heat consumption of cold wind intrusion:

$Q_6 = NQ_3 = 1 \times 121.4 = 121.4$ W

4. Ground division (64 m^2)

$R = R_0 + 10/0.04 = 0.47 + 250 (\text{m}^2 \cdot \text{K})/\text{W}$

$K = 0.004 \text{W}/(\text{m}^2 \cdot \text{K})$

$Q_7 = 0.004 \times 64 \times (18+7.5) = 6.5$ W

5. Thermal index:

We calculate the total load of the building based on the above

$Q_{\text{total}} = Q_1 + Q_2 + Q_3 + Q_4 + Q_5 + Q_6 + Q_7 = 74.6 + 87.8 + 121.4 + 36.9 + 48 + 121.4 + 6.5 = 496.6$ W

The building heat index is $Q_f = 496.6/36 = 13.8$ W/m^2

6.1.2 Design of the system

The final used design description is presented in the following. The configurations of PV/T collector are shown in Figure 6-2. The inclined angle of the collector is 47°, and it faces south. Based on the calculation results for this office room's thermal index, to satisfy the heating load demand for this room, the total surface area for PV/T collector with two arrays is designed for 2.6 m^2, and that for each one is 1.3 m^2. The structure of the collector is shown in Figure 6-2 (b). From top to bottom, the collector is composed of the metal frame, glass, two EVA files, and the film consists of the solar cell, the absorber, and the back insulation.

Figure 6-3 also shows the geometrical parameters for the collector. The thermal absorber consists of two metal layers and the thickness is 4 mm. The front metal is flat, while the back metal has evenly-distributed extruded pin-fins. These fins increase the heat exchange area and further strengthen the fluid disturbance. Stronger flow disturbance enhances the heat transfer coefficient. For further strengthening of the fluid disturbance, we design two types of pin-fins, which have different sizes. The shapes of the two pins are both cylindrical. The diameters of the pins are 3 cm and 1.5 cm, respectively.

There are two water inlets and water outlets which are located at the bottom and top of the absorber, respectively. The dimension of an absorber is 123 cm × 90 cm. The distance between the two neighbor fins is 12.5 cm. For installation, the distance from the edge to the absorber is 2 cm for the top, and 0.5 cm for the bottom. The distance between the pin and the edge is 5.5 cm. For installation, there is one blank area of 27 cm × 23 cm at the top, and the installation angle is 47°, south-facing. As the solar angle is near 50 °N at the noon and considering the seasonal changes, especially in the heating season the solar angle is 26.5 °N. To gain the most solar energy, the solar angel 47° is chosen.

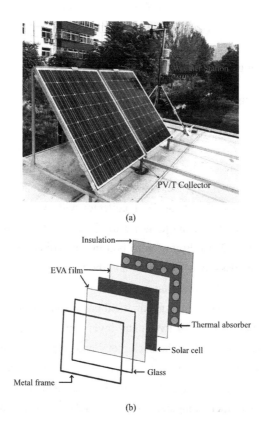

Figure 6-2 Configuration of PV/T collector: (a) photo; and (b) schematic of the structure

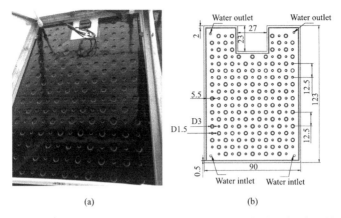

Figure 6-3 Configuration of the novel solar ultra-thin superconducting absorber: (a) photo; and (b) schematic (cm)

The schematic of the experimental setup and the working principle of PV/T system are shown in Figure 6-4. Figure 6-5 further shows the pictures of the main equipment, i.e., PV/T collector, the water tank and the evaporator of ASHP system in the experiment. In this experiment, PV/T system and ASHP system are combined in parallel mode. ASHP system has two functions, including gaining the energy from the outside air and gaining energy through heat exchange with the water tank which stores the heat gained by PV/T collector.

PV/T system gains solar energy and transfers it into electricity and heat. The electricity produced by PV /T collector is stored in the battery and supplied to the users. The electric box, as shown in Figure 6-5, is used to measure the total amount of electrical energy. The heat energy is supplied to the water tank for storage, which is supplied to ASHP system for room heating and hot water supply.

The design parameters of the main equipment in PV/T-ASHP system are shown in Table 6-1. The water tank volume is 100 L, which is used for hot water storage. The refrigerant used in ASHP system is R 22. The experiment is located in an office room near Beijing (117°E, 40°N), a cold region of China. The lab room is 3.5 m × 3 m × 3 m, the door is 2 m × 0.85 m and the window is 1.1 m × 1.1 m, south facing. This system is operated according to the office room working schedule, it is turned on at 8 am and turned off at 5 pm seven days a week.

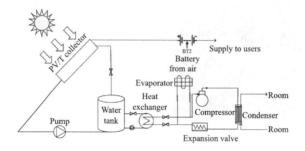

Figure 6-4 The schematic of the experimental setup of PV/T-ASHP system

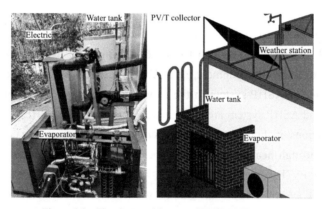

Figure 6-5 The picture of the part of PV/T-ASHP system

Table 6-1 The parameters of the main equipment

Devices	Number	Size/Capacity
PV/T collector	2	1.3m × 1m
Water tank	1	100 L
Pump	2	50 L / min
Condenser	1	3,500 W
Heat exchanger	1	3,000 W
Evaporator	1	2,500 W
Expansion valve	1	1.8 MPa / 0.5 MPa

6.1.3 Measurement and calculation

The external environment conditions are measured by the weather station installed next to PV/T collector [Figure 6-2 (a)]. The measurements include the external temperature, humidity, wind speed and solar irradiation. For PV/T-ASHP system, the main measurements include the water temperature at the solar collector inlet, the temperature of the outlet water from the tank and that at the thermal absorber surface temperature. We use the temperature probe with the accuracy of ±0.01 ℃ for the measurement. The water flow use an ultrasonic heat meter, with the range of 0.07-7 m³/h and the accuracy of level-2, ±0.14 m³/h. Experiments were operated from 9:00 to 17:00, and the time interval between the measurements was 10 min.

The performance of this PV/T-ASHP system is described separately, for PV/T collector system and ASHP system, mainly include the COP of ASHP system. The performance indicators are the COP of ASHP system, the thermal efficiency of PV/T collector, and the self-sufficiency of electricity and thermal energy for PV/T-ASHP system. These indicators are defined as follows:

$$\text{COP} = \frac{Q_{user}}{P_{ASHP}} \qquad (6\text{-}1)$$

$$\eta = \frac{Q_{gain}}{S} \qquad (6\text{-}2)$$

$$R_{ele} = \frac{P_{PV/T}}{P_{sys}} \qquad (6\text{-}3)$$

$$R_{heating} = \frac{Q_{gain}}{Q_{user}} \qquad (6\text{-}4)$$

where, COP is the coefficient of the ASHP performance, Q_{user} is the energy capacity supplied to users, P_{ASHP} is the power consumption by ASHP system. η is the thermal efficiency of PV/T collector; S is the total solar irradiation gained by PV/T collector, and Q_{gain} is the heat transfer capacity, which is stored in the water tank. Furthermore, R_{ele} represents the self-sufficiency of electricity, $P_{PV/T}$ is the electric power produced by PV/T collector, P_{sys} is the power consumption of the system, and $R_{heating}$ is the self-sufficiency of heat during heating seasons.

6.1.4 Conclusion

In this chapter, the model of the proposed novel ultra-thin superconducting solar absorber in PV/T collector is built through CFD software. Then the novel absorber is simulated under different boundary conditions and its performance for different constructions of the absorbers is compared. Finally, the performance under different conditions is analyzed.

Several main conclusion includes: (1) The heat flux has a significant impact on the efficiency (η) of the ultra-thin superconducting absorber. Higher heat flux enhances the solar heat transfer of the absorber hence significantly increases the efficiency (η) from 10.8% to near 60%. (2) From the result, the impact of the inlet temperature can not significantly affect the performance of the absorber and by increasing the inlet water temperature, the difference between the inlet and the outlet temperatures is reduced and the efficiency of the ultra-thin superconducting absorber is reduced from 52.7% to 49%. (3) With the increase of the inlet water velocity from 0.2 m/s to 1 m/s, the efficiency is reduced from 70% to under 10%. For the inlet water velocity larger than 0.6 m/s, the temperature difference between the inlet and outlet water temperatures is near 0. (4) For the inlet water velocity of 0.5 m/s, the first type has the highest heat collecting (27.8×10^6 J/h) and efficiency (68%) amongst all types, except the all big type with efficiency (69%) but the difference is small. Moreover, the all big type needs more power in practice due to the larger resistance compared with the first type. The experiment performance of the novel ultra-thin superconducting absorber needs further study in the future.

6.2 Experiment of testing of the solar collector in PV/T-ASHP system and model validation using Matlab

One of the most common methods to improve the thermal performance of the collector is by optimizing its design. The common designs include a sheet-and-tube structure, rectangular tunnel with or without fins/grooves, flat plate tube, micro-channel heat pipe array/ heat mat, extruded heat exchanger, roll-bond heat exchanger, and cotton wick structur. [1] Several studies on the thermal absorber have been conducted over the past few decades. A sheet-and-tube absorber requires a relatively low investment as the corresponding industry is rather mature and it also provides satisfactory performance. [2] Nevertheless, its structure is complex. The application of sheet-and-tube absorber is limited because of its large weight. [3]

Amongst the existing structures, the absorber of a rectangular tunnel with or without fins/grooves has the lowest weight and incurs the lowest investment. [4-5] However, its heat transfer efficiency is the lowest. [6-7] Although the contact between PV cells and absorber of a flat plate tube is simple, the thermal resistance of its structure is high. [8-9] An absorber with a micro-channel heat pipe, also called a heat mat, improves thermal performance. However, the thermal resistance is also increased. [10-12] The applications of the roll-bond absorber and cotton wick structure absorbers are also limited [13] mainly because of their high construction and maintenance cost. The extruded absorber demonstrates high potential compared to the above absorbers, particularly, the super-thin extruded pin-fins absorber. It generally achieves higher efficiency than other designs of the absorber plates. This is because its special design enhances the heat transfer and its weight and the investment required are significantly lower than the extruder absorber. [14-15] The main results are as follows: the hybrid PV/T panel enhances the electrical return of PV panels by nearly 3.5% and increases the overall energy output by nearly 324.3% under certain circumstances. Their results demonstrate that PV/T panel achieves an electrical efficiency of approximately 16.8% (5% improvement compared with the stand-alone PV panel), and yields an additional amount of heat with a thermal efficiency of nearly 65%.

However, there are two issues. One is that the existing studies are either solely based on simulations, or conducted in laboratory conditions .[14-15] Hence, the feasibility of its actual marketable applications has not been verified. In this study, our objective is to investigate the application performance of a novel construction extruded absorber. The effect of its application on the thermal performance of an entire PV/T-assisted ASHP system is studied and analyzed for an office room during the heating season.

Moreover, the effects of several uncertain factors such as the dynamic weather conditions on this system have not been studied.[28] In this chapter, the variation in the thermal efficiency of the solar collector under different environmental conditions is also analyzed using both mathematical and experimental methods. This study addresses the research gap in the related literature and is essential for future market release.

6.2.1 Working principle of the solar collector

PV/T collector absorbs solar energy and produces electricity which is stored in a battery. It also produces heat, which is transferred to the absorber. The absorber supplies the heat to other equipment such as the water tank for further use. The PV cell absorbs part of the solar energy and transforms it into electricity. The remaining solar energy is transformed into heat which is absorbed by a thermal absorber.

The collector can be divided into multiple layers, as shown in Figure 6-6. Three hypotheses are considered to simplify the model. Firstly, the temperature at each layer is assumed to be uniformly distributed. Secondly, the glass absorbs the solar radiation, where the solar energy is partly absorbed and the remaining is reflected. Thirdly, considering that the EVA film is very thin and R_{PV} is marginal, in our modeling, we ignore the heat loss that happens through the PV cell to the surrounding environment.

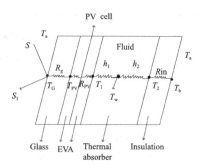

Figure 6-6 Thermal resistance network of PV/T collector

Chapter 6
Experiment Study of PV/T-ASHP System in an Office Room Application

For the glass surface, part of the solar radiation (S) on the surface/glass of the collector is reflected in the atmosphere, and the remaining part is absorbed and transferred to the PV-cell layer. Equation (6-5) is developed to show the mathematical relationship among different parts of energy.

$$S - S_1 + h_G (T_G - T_a) = (T_{PV} - T_G) / R_g \quad (6\text{-}5)$$

where, S represents the solar radiation that reaches the collector glass surface, S_1 is the solar radiation that reflects to the atmosphere from the glass surface, h_G is the glass radiative heat transfer coefficient (J/h · m^2 · K), T_a is the ambient temperature (K), T_G is the glass surface temperature (K), T_{PV} is the PV cell surface temperature (K), and R_g is the thermal resistance of the glass layer (h · m^2 · K/J).

Part of the solar energy that is absorbed by the PV cell is then transformed into electricity and stored in the battery. The remaining energy is transferred to the thermal absorber. The heat transfer is calculated using the following Equation (6-6).

$$\frac{(T_{PV} - T_G)}{R_g} = \frac{(T_1 - T_{PV})}{R_{PV}} \quad (6\text{-}6)$$

where, T_1 is the thermal absorber surface temperature (K) and R_{PV} is the thermal resistance of the PV cell (h · m^2 · K/kJ).

In the thermal absorber, the water absorbs heat and then supplies the energy to the users. The energy balance is as the following:

$$q_u = h_1 (T_1 - T_w) - h_2 (T_w - T_2) \quad (6\text{-}7)$$

where, q_u is the heat absorbed by the water (J/h · m^2), T_w is the mean water temperature in the thermal absorber (K), T_2 is the back surface temperature of the thermal absorber (K), h_1 is the heat transfer coefficient to the channel at the front, and h_2 is the heat transfer coefficient to the channel at the back (J/h · m^2 · K).

The heat is then transferred to the back-insulation material according to the following equation:

$$\frac{T_2 - T_b}{R_{in}} = h_b (T_b - T_a) + Q_{loss} \tag{6-8}$$

where, T_b is the back surface temperature of the insulation (K), R_{in} is the thermal resistance of the insulation (h · m² · K/J), h_b is the insulation radiative heat transfer coefficient (kJ/h · m² · K), and Q_{loss} is the heat loss to the surrounding environment through insulation (kJ/h · m²). The convection coefficient is $h_{loss-c} = \frac{8.6V^{0.6}}{L^{0.4}}$, where, V is the ambient wind speed. Radiation heat transfer coefficient is $h_{loss-r} = \varepsilon\sigma (T_G + T_S)(T_G^2 + T_S^2)$, where, the sky temperature is $T_S = 0.055T_a^{1.5}$. [16]

The absorber thermal efficiency is essential for evaluating the performance of PV/T collector. In this experiment, the solar collector thermal efficiency is calculated with Equation (6-9):

$$\eta_{th} = \frac{Q_u}{IAt} 100 \tag{6-9}$$

where, η_{th} is the thermal absorber heat efficiency (%), Q_u is the heat of the water absorbed through thermal absorber (kJ), I is the solar irradiation absorbed by the collector (W/m²), and A is the collector area (m²), t is the time over which the energy is collected or the duration of the experiment (s).

The heat transfer efficiency is evaluated via both experimental measurement and mathematical model calculation. According to Emmanouil et al. [17], the error in the heat loss calculation includes the measuring set-up error (1.5%), imperfections of the energy model (2.6%), and variation in the meteorological conditions (3.5%). The total combined uncertainty is 9% for a confidence level of 95% and a coverage coefficient (k) of 2.

The external environmental conditions are measured by the weather station installed next to PV/T collector. The measurements include the external temperature, humidity, wind speed, and solar radiation. The main measurements for PV/T-ASHP system include the water temperature at the solar collector's inlet, temperature of the outlet water from the tank, and temperature of the thermal absorber's surface. We use a temperature probe with an accuracy of ±0.01 °C for our measurement. The water flow is measured using an ultrasonic heat meter with a range of 0.07–7 m^3/h and level-2 accuracy (±0.14 m^3/h). Experiments are conducted during 9:00–17:00. And the time interval between the measurements is 10 min.

6.2.2 Matlab model of the solar collector

In this section, the mathematical model of the solar collector for calculating its thermal performance is presented. To simplify the model, we make the following assumptions: (1) There is no heat loss through the back-insulation material and the collector edge. (2) The heat transfer coefficient is constant and uniform for the pin-fins. (3) There is no heat loss during the process of the heat transfer among the different layers in the solar collector. (4) The temperature gradient across the solar collector thickness is omitted as in.[15] Furthermore, the detailed calculation for the solar collector is as the following.

1. Heat transfer for the novel thermal absorber with frustum pin-fin banks

Based on the above hypotheses, the heat transfer coefficient is constant and uniform throughout the fin surface. The transfer process is a one-dimensional steady-state heat conduction problem with variable cross-section. Because the diameter and the distance are linear, the length of the frustum H (4 mm)<D. D is the diameter of the frustum. The frustum pin-fin is shown in Figure 6-7. We further assume that the heat transfer coefficient is constant and uniform over the whole fin surface.

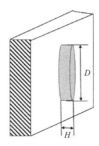

Figure 6-7　The schematic of the simplified frustum pin-fin

Considering the big pin-fin as an example, the fin efficiency is:

$$\eta_f = \frac{\text{th}(mH')}{mH'} \qquad (6\text{-}10)$$

where, the variable m is:

$$m = \sqrt{\frac{4h}{\lambda d}} \qquad (6\text{-}11)$$

and H' is: $H' = H + \dfrac{D_{big}}{4}$

and the small pin-fin is the same as the big pin-fin, except for its diameter.

Under steady conditions, the useful heat gain Q_u is equal to the heat conduction through both pin-fin and un-finned surfaces, which is calculated as:

$$Q_u = \int Aq_u \, dA = h_{conv,fT}(\eta_f A_{fin} + A_{unfin})(T_1 - T_f) \qquad (6\text{-}12)$$

$$T_f = (T_{in} + T_{out})/2 \qquad (6\text{-}13)$$

where, A_{fin} is the total surface of all the fins include big fin and small fin, the A_{unfin} is the area of the un-finned of the plate surface,

$$A_{fin} = \frac{n}{2}\pi(D_{big} + D_{small})H \qquad (6\text{-}14)$$

$$A_{unfin} = A_p - \frac{n}{2}\pi\frac{D_{big}^2 + D_{small}^2}{4} \qquad (6\text{-}15)$$

And n is the total number of pin-fins.

For the fully developed turbulent flow in the channel, the Nusselt number is calculated by Gnielinski Equation [25]:

$$Nu_f = \frac{(f/8)(Re - 1000)Pr_f}{1 + 12.7\sqrt{f/8}(Pr_f^{2/3} - 1)}[1 + (\frac{d}{l})^{2/3}]C_t \qquad (6\text{-}16)$$

Where, l is the length of channel and f is the Darcy coefficient of the turbulent flow, $f = (1.82lgRe - 1.64)^{-2}$.[18]

2. Heat convection across the line pin-fin banks

The aligned arrangement pin-fin banks are shown in Figure 6-8. For this novel thermal absorber, the distance between fins L_1 is equal to L_2. Considering the difference of the diameter between the big and small fins, we take the mean value as the diameter of the fin.

$$D_{mean} = \frac{D_{big} + D_{small}}{4} \qquad (6-17)$$

In this experiment, the number of rows of the aligned pin-fin banks is 18, over 16. The Nusselt numbers are calculated by Zhukauskas equations, shown in Table 6-2.

Then the useful heat is gained by the flow convection Q_c.

$$Q_c = m_w C_p (T_{out} - T_{in}) = hA (T_{out} - T_{in}) \qquad (6-18)$$

$$h = \lambda N_u / D_{mean} \qquad (6-19)$$

The mass flow rate of water m_w is a design parameter in the experiment.

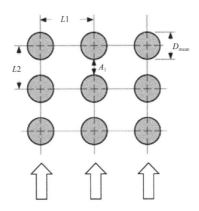

Figure 6-8 Arrangement of the aligned pin-fin banks

Table 6–2 Calculation equations for determining the average surface heat transfer coefficient of the fluid across the aligned pin-fin banks [18]

Equation	Re
$Nu_f = 0.9 Re_f^{0.4} Pr_f^{0.36} (Pr_f/Pr_w)^{0.25}$	$1 - 10^2$
$Nu_f = 0.52 Re_f^{0.5} Pr_f^{0.36} (Pr_f/Pr_w)^{0.25}$	$10^2 - 10^3$
$Nu_f = 0.27 Re_f^{0.63} Pr_f^{0.36} (Pr_f/Pr_w)^{0.25}$	$10^3 - 2 \times 10^5$
$Nu_f = 0.033 Re_f^{0.8} Pr_f^{0.36} (Pr_f/Pr_w)^{0.25}$	$2 \times 10^5 - 2 \times 10^6$

3. The simulation steps

Based on Equations (5)-(19), a computation program is developed for simulation and its flowchart is shown in Figure 6-9. The initial operation condition is as the following. The base setting is under 800 W/m², wind speed is 1 m/s, and the outside temperature is 25°C, the inlet water temperature is 20°C, and the inlet water velocity is 0.5 m/s.

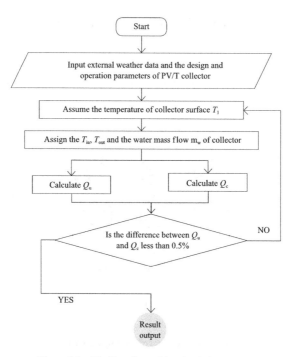

Figure 6-9 The flowchart of the simulation process

As illustrated in Figure 6-9, the computation steps are as the following: (1) Input the external weather conditions and the design parameters. (2) Set the operating conditions. Assume the absorber surface temperatures T_1. (3) Assign the experimental operation parameters concerning water. These include the water mass flow m_w, and the inlet temperature T_{in} and outlet temperature of the absorber T_{out}. (4) Calculate the heat gain through heat conduction. (5) Calculate the heat gain from the water convection part. (6) Compute the difference between Q_u and Q_c. If $|(Q_u - Q_c)/Q_c| \leq 0.5\%$, then calculate the thermal efficiency of the absorber and then output. Otherwise, reset the absorber surface temperatures T_1 and T_2, and repeat the previous steps.

6.2.3 Validation of the Matlab model

To verify the accuracy of the mathematic model, the model results are compared with the corresponding experimental results for different external environmental conditions. The comparison results are shown in Table 6-3. As it is seen that the largest error is 9.6% and the average error is 1.5%. These results confirm the model's accuracy and validity.

Considering the uncertainty of during the measurement, these errors are reasonable for engineering research. Note that a few assumptions are made in constructing the simulation model, e.g., the heat loss through the frame to the surrounding is ignored. Errors arise during the measurement and the environmental conditions are varying over time, which results in inaccuracy. For example, although the wind speed is close to 1 m/s, it can not be maintained at 1 m/s. Therefore, the largest error of 9.55% between the simulation and the experiment is reasonable. [14]

Table 6-3 Comparison of simulations and experiments

S (W/m^2)	T_a (℃)	V_a (m/s)	η (%)		Error (%)
			Sim	Exp	
586	−5	3	12	12.9	6.98
612	0	1.5	23	22.5	2.31

Continued

S (W/m^2)	T_a (°C)	V_a (m/s)	η (%)		Error (%)
			Sim	Exp	
602	5	2	23	23.7	3.10
606	10	3	24	23.7	1.07
598	15	1.5	31	32.4	4.39
590	20	0.5	40.2	41.1	2.19
696	25	1	39.8	39.8	0.06
601	30	1.5	40.1	40.0	0.25
323	10	0.5	20.5	22.4	8.48
410	10	0.5	28.2	29.2	3.42
495	10	0.5	31.2	32.2	3.11
656	10	3	24.1	22.0	9.55
679	10	0.5	36.2	34.6	4.62
789	10	1.5	30.2	31.0	2.58
924	10	2	32.1	31.8	0.94

6.2.4 Parametric study of the solar collector

The experimental results on the performance of the solar collector are presented and analyzed. Moreover, the application results and simulation results are analyzed and compared in this section. This is to investigate the variations in the thermal efficiency of the novel solar collector under different environmental conditions.

Compared with the extensive variations in solar radiation and ambient temperature, the variation in the wind speed is generally smooth. We selected sunny days during the experimental period. The thermal efficiency is shown in Figure 6-10. This figure shows the thermal efficiency of the solar collector versus the solar irradiance and outdoor temperature. The efficiency of the solar collector varies from 5% to nearly 60% during the period of experiment. Meanwhile, the average efficiency is approximately 30.7%. Figure 6-10 shows that the thermal efficiency of the solar collector is increased by increasing solar radiation. Most of the values vary from 20% to 42.5%. For the ambient temperature of approximately 26°C, the solar radiation is approximately 600 W/m², the wind speed varies from 1.5 m/s to 2 m/s, and the thermal efficiency is at its highest (65%). Figure 6-10 also indicates that the change for the thermal efficiency along with the environmental conditions is rather complex. As shown in the right corner in Figure 6-10, the thermal efficiency is high because the wind speed for this figure is not stable. This may be caused by multi-environment factors including solar irradiation, ambient temperature, and wind speed.

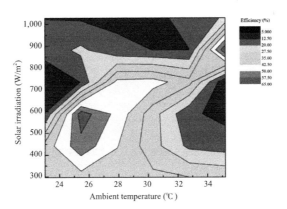

Figure 6-10   Thermal efficiency of the PV/T solar collector (where the wind speed < 3 m/s)

Figure 6-10 suggests that the thermal efficiency of the solar collector is affected by a few factors. We further consider a certain day, October 3rd, 2018, as an example. The reason for choosing this day is that it was a sunny day which was common during 2018. Moreover, the variation of the environmental conditions was more stable than the rainy days.

Figure 6-11 (a) and (b) show that the variations in solar radiation and ambient temperature follow particular trends, whereas the thermal efficiency and wind speed fluctuate without following an apparent trend. We set the model to clarify the influence of each factor (including the environment and the collector operation) on the thermal efficiency. The results are presented in the following section.

ASHP Assisted Solar Energy Systems:
Classification, Applications, and Performance Research

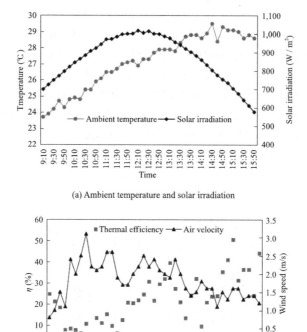

(a) Ambient temperature and solar irradiation

(b) Wind speed and thermal efficiency

Figure 6-11　The results of a typical date

6.2.5　Comparison of the experiment and the Matlab model

Using the mathematical model, we calculate the thermal efficiency of the solar collector under different external environmental and operational conditions. To identify the effect of these factors on the thermal efficiency, we set benchmarks for the external environmental conditions, i.e., solar radiation of 600 W/m^2, ambient the air temperature of 10°C, outdoor wind speed of 1 m/s, inlet water temperature of 10°C, and inlet water velocity of 0.3 m/s. [8] The operating conditions of the system are also constants, i.e., the installation angle of the solar collector (47°). The results on the thermal efficiency of the solar collector are discussed in the following.

To identify the impact of solar radiation and wind speed on the thermal efficiency, the solar radiation is varied from 300 W/m² to 1000 W/m² at intervals of 100 W/m², and the wind speed varies from 0.5 m/s to 3 m/s, while other environmental conditions remain unchanged. It can be seen from Figure 6-12 that the thermal efficiency (η) increases by increasing the solar radiation. Higher solar radiation enhances heat transfer by increasing solar energy gain. The thermal efficiency of the solar collector is increased from approximately 18% to nearly 40%. Moreover, Figure 6-12 shows that thermal efficiency is increased by increasing solar radiation, whereas it is decreased by increasing the wind speed. Comparison of the simulations and experimental results shows that the values are highly similar and the average error is 4.67%.

It is further seen that the thermal efficiency is increased by increasing the ambient temperature, whereas it is decreased by increasing the wind speed as in Figure 6-13. The thermal efficiency increases by increasing the ambient temperature because a higher ambient temperature improves the solar gain of the collector. Besides, as the temperature difference between the solar collector and the environment is reduced, the heat loss to the surrounding environment is also reduced. The experimental results illustrate that the thermal efficiency increases from 24% to 46% when the outside air temperature varies from −5°C to 30°C. The average error between the experimental and simulation results is 2.54%. The largest error is observed to be 6.98% when the ambient temperature is −5°C and wind speed is 3m/s.

In general, the wind speed varies more than the solar radiation level and ambient temperature. To obtain more accurate results, the influence of the wind speed on the thermal efficiency of the PV/T solar collector is analyzed via simulation. The simulation results are shown in Figure 6-12. It displays the variation trend upon the outside wind speed variations from 0 m/s to 5 m/s while other parameters are constant, i.e., solar radiation of 600 W/m² and the ambient temperature of 10 °C. Figure 6-13 shows that thermal efficiency is decreased by increasing the outside wind speed. This is because the higher the wind speed, the higher the heat transfer coefficient between the environment and the solar collector. Therefore, the heat loss to the environment from the solar collector is increased. Based on these results, the thermal efficiency is decreased from 37.5% to nearly 22% by increasing the outside wind speed. The thermal efficiency is generally decreased for the wind speed lower than 1 m/s. Meanwhile, the thermal efficiency is reduced more rapidly where the wind speed varies from 1 m/s to 3 m/s. For the outside wind speed higher than 3 m/s, which is an extreme condition, the thermal efficiency varies less.

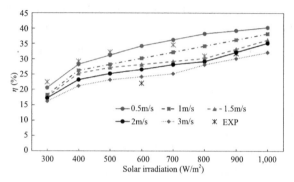

Figure 6-12　Thermal efficiency under different solar irradiation and wind speed where the ambient temperature is 10°C

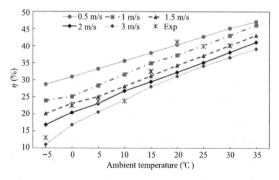

Figure 6-13　Thermal efficiency under different ambient temperatures and wind speeds, where the solar radiation is 600 W/m²

6.2.6　Conclusion

An experimental setup of PV/T-ASHP system for an office room is constructed in this section to verify the actual application performance of the proposed ultra-thin superconducting solar collector with pin-fins. The heating season performance of this system is also analyzed. Moreover, the thermal performance of the solar collector of PV/T with pin-fins under different environmental conditions is investigated through both simulations and experimental methods to clarify the variation in the thermal efficiency of the solar collector in different environmental conditions. Three environmental factors are analyzed: solar radiation, ambient temperature, and wind speed. The absorber thermal efficiency is calculated and compared between the simulation and experiment. The main conclusions are as follows:

A parametric study of the thermal efficiency of the solar collector under different environmental and operating conditions reveals a good agreement between the simulation and experimental results. The average error between the simulation and experiment is 1.52% which is acceptable because there are influencing factors both in the simulation and experiment. The higher the solar radiation and ambient temperature, the higher the thermal efficiency. Meanwhile, the higher the wind speed, inlet water temperature, and inlet water velocity, the lower the thermal efficiency. The highest efficiency of this solar collector is nearly 65% during the experiment.

The results presented in this section are anticipated to be useful for further design of PV/T system. Further work is needed to optimize the design and operation for enhancing the system's performance. Further research should also be conducted on this system to achieve more benefits in terms of environment-friendliness and energy efficiency.

6.3 PV/T-ASHP system performance

For PV/T-ASHP system, few papers have done the experiment research, many authors just pay attention to the simulation as the system is complex and difficult to set up in laboratory. And most of the researches focused on the energetic performance. For example, Wang et al. [19] built experiment of PV/T-ASHP hot water system in laboratory and tested the system. The system was comprised of independently developed flat plate solar PV/T collector based on micro-channel heat pipe array and air source heat pump, which were combined by a new composite evaporator. The COP of heat pump reduced from 5.61 to 1.69 and the average was 3.03. Comprehensive COPsys of PV/T-air composite heat pump system ranged from 6.07 to 1.33, the average was 2.99. Raghad et al. [20] designed BIPV/T-ASHP system on a test hub and found that the seasonal COP could be increased from 2.74 to a maximum value of 3.45 for direct coupling of BIPV/T+ASHP without the use of diurnal thermal storage. The heat pump electricity consumption is reduced by 20% for winter. PV/T-ASHP system are usually composed by three parts: (1) PV/T collector or BIPV/T system on facade; (2) heat pump unit; and (3) thermal energy storage (TES) system, including insulated concrete forms walls, ventilated concrete slab, gravel/sand bed and water tank storage. Qu et al. [21] examined a novel solar photovoltaic/thermal integrated dual source heat pump water heating system at different ambient temperature. Besagni et al. [22] focused on the solar-assisted heat pumps for heating and cooling to produce DHW. The system was coupled with PV or PV/T system. They made the examination under different ambient temperature, system parameters and operating modes to compare the performance of the system, mainly the COP.

One special condition facing by ASHP system in extreme condition is frosting. To avoid the deterioration of heat transfer effect and performance drop of unit due to the evaporator frosting in low temperature condition with the circulated air heated by the air collector, Cao et al. [29] and Li et al. [23] proposed a PV/T-ASHP integration heating system, which aimed to operate safely, stably and efficiently in winter in cold region. PV/T air collectors were arrayed on the building envelop composing an integrated PV/T curtain wall. In PV/T, up and down sides are respectively opened outlets and inlets connected to ASHP unit with air ducts forming a forced circulation in the system. Besides, they assumed that there is almost no overheating during the energy transfer process, freezing, boiling, corrosion and leakage problems. Lu et al. [24] studied the performance of PV/T-AHP system in winter. To improve the performance, they added vapor injection to this system. As a result, the COP was 3.45 when the ambient temperature average was −1.13 °C and solar irradiation was 164.03 W/m^2.

6.4 Performance study of PV/T-ASHP system

Most of the researches related to PV/T-HP system were focused on their performance. However, almost all of them were conducted using simulation or in laboratory experiments, where the experimental conditions and duration were limited. Hence, there is a lack of experimental investigation throughout the year, which is essential for marketing. Besides, the economic and environmental analyses in the existing researches are very limited. In this book, we present a PV/T assisted ASHP system that supplies heating and domestic hot water for users. We then implement our design in Beijing, China. A series of experiments were also conducted throughout 2018 to investigate the performance of this system. The results of the COP of ASHP system, the thermal and electrical efficiency of the PV/T collector, the self-sufficiency rate of thermal energy, and the self-sufficiency rate of electricity of this system in each month, are identified and presented in this Section. Moreover, the analysis of economic and environmental impacts, such as the payback time and pollution emission are presented in this section.

6.4.1 Description of the experimental device

The schematic of the experimental setup and the working principle of PV/T system are shown in Figure 6-14. The main equipment is also shown in Figure 6-15. Figure 6-16 shows the configuration of the novel solar absorber.The design parameters of the main equipment in PV/T-ASHP system are shown in Table 6-4. The water tank volume is 100L, which is used for hot water storage. The refrigerant used in the ASHP system is R22. The experiment is located in an office room in Beijing (117°E, 40°N), which is in the cold region of China. The office room is 3.5 m ×3 m ×3 m, the door is 2 m × 0.85 m and the window is 1.1 m × 1.1 m, south facing, a top layer office room. Stronger flow disturbance enhances the heat transfer coefficient for the PV/T collector. [25] To further strengthen the fluid disturbance, we design two types of pin-fins of different sizes. The shape of both of these pins is cylindrical and the diameters of the pins are 3 cm and 1.5 cm, respectively.

ASHP Assisted Solar Energy Systems:
Classification, Applications, and Performance Research

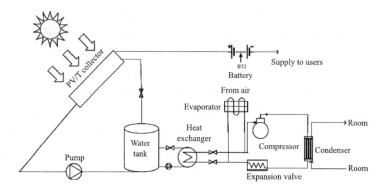

Figure 6-14　The schematic of the experimental setup of PV/T-ASHP system

Figure 6-15　The picture of the part of PV/T-ASHP system

Figure 6-16　Configuration of the novel solar absorber

Table 6-4 The parameters of the main equipment

Devices	Number	Size/Capacity
PV/T collector	2	1.3 m×1m
Water tank	1	100 L
Circulation pump	2	50 L/min
Condenser	1	3,500 W
Heat exchanger	1	3,000 W
Evaporator	1	2,500 W
Expansion valve	1	1.8 MPa/0.5 MPa

6.4.2 Working principle of PV/T-ASHP system

For the energy transfer process, in this experiment, PV/T and ASHP systems are combined in parallel mode. ASHP system can either absorb heat from the air outside or from the water tank which stores heat gained by the PV/T collector. PV/T system gains solar energy and transfers it into electricity and heat. The electricity produced by the PV /T collector is then stored in the battery and supplied to the users. The electric box, as shown in Figure 6-15, is used to measure the total amount of electrical energy. The heat energy is supplied to the water tank for storage which is then supplied to ASHP system for room heating and hot water supply.

During the experiment period, at part of spring and autumn, winter (heating season), PV/T-ASHP system supplies electricity, hot water, and heating to the users from 9:00 to 17:00. ASHP system firstly absorbs heat from the water tank. If the energy is not sufficient to satisfy the demand, the ASHP changes to absorb heat directly from the outside air. While for part of the spring and summer season (cooling season), PV/T-ASHP system supplies electricity, cooling, and hot water to the users, ASHP is used for cooling and PV/T is used for supplying electricity and hot water.

6.4.3 Calculation of PV/T-ASHP system performance

Based on the experimental data collected in 2018, the monthly performances, i.e., the COP of ASHP system, the thermal efficiency of the

PV/T collector and the self-sufficiency of electricity and thermal energy for whole PV/T-ASHP system, the PV/T collector system and ASHP system are separately calculated, as defined by using following equations.

$$COP = \frac{Q_{user}}{P_{ASHP}} \qquad (6\text{-}20)$$

$$\eta = \frac{Q_{gain}}{S} \qquad (6\text{-}21)$$

$$R_{ele} = \frac{P_{PV/T}}{P_{sys}} \qquad (6\text{-}22)$$

$$R_{heating} = \frac{Q_{gain}}{Q_{user}} \qquad (6\text{-}23)$$

where, COP is the coefficient of the ASHP performance, Q_{user} is the energy capacity supplied to users, P_{ASHP} is the power consumption by ASHP system, η is the thermal efficiency of PV/T collector, S is the total solar irradiation gained by the PV/T collector, and Q_{gain} is the heat transfer capacity, which is stored in the water tank. Furthermore, R_{ele} represents the self-sufficiency of electricity, $P_{PV/T}$ is the electric power produced by PV/T collector, P_{sys} is the power consumption of the system. $R_{heating}$ is the self-sufficiency of thermal energy during the heating season.

6.4.4 Analysis of the result

The results of each month are listed in Table 6-5. Results of indoor temperature show that this PV/T-ASHP system satisfies the user demand for indoor thermal comfort, as the temperature is warmer than the design value (18°C) in the heating seasons. Similarly, during the cooling seasons, the indoor temperature is lower than the design temperature (28°C).

Table 6-5 shows the 24-hour average value of the solar irradiation in each month. The lowest irradiation level is in January, and May has the highest irradiation level. This is because, during the summer, the solar irradiation is stronger, whereas it is often more wind and rain during the winter. Therefore, the amount of electricity production from the PV/T collector in May is the highest including the total electricity and the useful electricity. The average ambient temperature is the highest, 28.4°C, in July and lowest, −3.4°C, in December. The PV/T collector absorbs the largest amount of heat in April, 6,264 kW. For the PV/T collector, the thermal efficiency is varied from 8.82% (in December, when the solar irradiation and ambient temperature are both low) to 41.70%, and the average is 28.9%. The total amount of the produced electricity is 9,330 kWh for the whole year, while the useful amount of electricity is 6,819.5 kWh throughout the year. The heat transferred by the solar collector is also 46,354 kW during the year. For the ASHP part, the average COP is ranged from 2.3 to 4.6, the average is 3.425 which is bigger than reported in. [19] The heating and cooling supplied to the office are equivalent to 36,115.5 W and the electricity consumption by ASHP was 10,432 kWh throughout the year.

Note that in January, the system only works for the first two weeks. Therefore, the average ambient temperature is a little higher than the average value over the month. Moreover, the office is not in use during January so the indoor temperature is 18°C, while the system is still on to avoid frozen. Therefore, the heating self-sufficiency rate in January is much higher than that in December.

For PV/T-ASHP system, the electricity self-sufficiency reaches the highest which is 78.61%, and during heating seasons, the thermal self-sufficiency reaches up to 88.42%. Even during December, it could cover the heat demand of 30.73% by the solar collector. While in cooling seasons, the heat absorbed by PV/T collector is only used for hot water. This suggests that this system can offer over half of the total energy consumption.

To analyze the real-time performance of this PV/T-ASHP system, we use a set of experimental data under typical climate conditions in different seasons (April 1st, 2018, June 30th, 2018, October 30th, 2018, December 31th, 2018). We consider these four days because they represent different climates from different seasons. Using this data we can draw general conclusions for similar days throughout the year, see the results in Figures 6-17 to 6-20.

For the windy day of April 1st, 2018, the results of this system are presented in Figure 6-17. This system supplied heating, electricity and hot

ASHP Assisted Solar Energy Systems: Classification, Applications, and Performance Research

Table 6-5 Experimental data obtained in 2018 for each month

Month	Average Solar Irradiation (W/m²·h)	Average Ambient Temperature (°C)	Average Humidity (%)	PV/T Total electricity (kWh)	PV/T Useful electricity (kWh)	PV/T Heat transfer (kW)	PV/T Thermal Efficiency (η)	PV/T Heating/cooling (kWh)	ASHP Electricity Consumption (kWh)	ASHP COP	Average Indoor Temperature (°C)	Self-sufficiency rate of electricity	Self-sufficiency rate of thermal energy
January (two weeks)	189	-1	28.3	734	525	2,554	21.66%	3,557.22	889	4	18	59.06%	71.80%
February	200.7	-0.6	30.3	684	460	3,269	26.11%	3,790.94	876	4.3	25	52.51%	68.23%
March	240.7	9.9	33	869	647	6,264	41.70%	3,794.78	823	4.6	20	78.61%	—
April	189.2	14.5	47.5	830	617	4,754.4	40.31%	2,745.25	798	3.4	21.8	77.32%	—
May	248.5	23.3	37.9	875	652.5	4,404.48	28.40%	2,027.9	835	2.4	25.7	78.14%	—
June	219.8	26.9	52.4	866	624	4,356.96	31.77%	2,196.45	856	2.6	30.2	72.90%	—
July	203.3	28.4	67.8	821	596	4,595.52	36.22%	2,121.77	800	2.7	27.8	74.50%	—
Aug	207.6	26.8	62.4	755	585	4,352.64	33.59%	2,262.92	986	2.3	27.4	59.33%	—
Sep	217.6	24.5	59.2	794	575	4,813.44	35.45%	2,774.78	942	2.9	26.4	61.04%	—
Oct	217.2	13.5	57.3	717	518	2,981.28	22.00%	3,371.62	899	3.8	22.5	57.62%	88.42%
Nov	217.2	5.6	32	702	512	2,812.3	20.75%	3,580.02	912	3.9	20.3	56.14%	78.56%
Dec	217.4	-3.4	19.1	683	508	1,196	8.82%	3,891.83	916	4.2	18.5	62.25%	30.73%
Total				9330	6819.5	4,6354.0		3,6115.5	1,0432				

water. The average indoor temperature was 20°C. As seen in Figure 6-17 (a), the temperature was decreased from 11°C to 4°C during the daytime. The solar irradiation during the daytime was increased from 9:00 to 11:00, then it fluctuated slightly from 11:00 to 15:50, and then quickly reduced. The values varied from the lowest, 150 W/m^2 in the morning at 9:00, to the highest, 900W/m^2 at 15:40. For the PV/T collector, Figure 6-17 (b) shows that at the beginning, its absorber surface temperature is the same as the outlet water temperature, 16°C. Then as the collector absorbed heat from the solar, its surface temperature was increased up to nearly 40°C. The thermal efficiency of the collector varied from the lowest 10% to the highest 60%. This is because it is affected by the external environment conditions including the ambient temperature and solar irradiation. The thermal efficiency was also varied a lot from 9:00 to 11:20, where the ambient temperature and solar irradiation varied frequently. In Figure 6-17 (c), the photoelectric efficiency is shown to reduce from 11.8% to 2.5%, with an average of 4% on April 1st, 2018. The COP of ASHP is increased from 2 at 9:00, to its largest, 3.6 at noon. It is then reduced to 2.5 in 17:00.

For an overcast day on June 30th, 2019, the system produced both cooling and electricity. The performance results are shown in Figure 6-18. The average indoor temperature is 28.6 °C. Figure 6-18 (a) shows the variation of the solar irradiation, especially during 12:00 to 15:00, with the lowest value around 200 W/m^2 at this period. The ambient temperature had a slight increase from 28°C to 33°C. Figure 6-18 (b) shows that the absorber surface temperature and the outlet water temperature of the collector were increased to around 10°C during this day. The thermal efficiency of the collector also varied a lot from 12:00 to 15:00, from 18% to 50%, and the average thermal efficiency was around 30%. The variation trend of thermal efficiency was caused by the significant influence of solar irradiation. In Figure 6-18 (c), the photoelectric efficiency also varied a lot from 12:00 to 15:00, due to the significant impact of the solar irradiation and the average photoelectric efficiency was about 3.8%. For the ASHP, the COP had an upward trend, from 2.3 at the beginning, i.e., 9:00, to the highest value, 3.5 at 14:00.

ASHP Assisted Solar Energy Systems:
Classification, Applications, and Performance Research

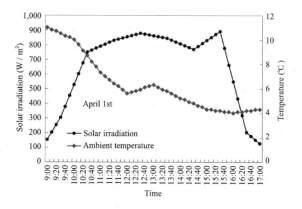

(a) The environmental conditions

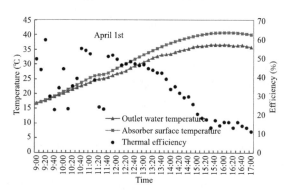

(b) The performance of the PV/T collector

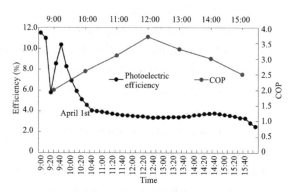

(c) The performance of the collector and ASHP

Figure 6-17　The results of a typical day (April 1st)

Chapter 6
Experiment Study of PV/T-ASHP System in an Office Room Application

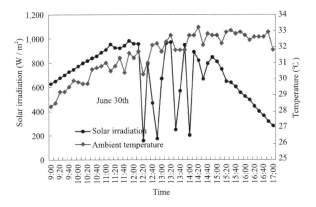

(a) The environmental conditions

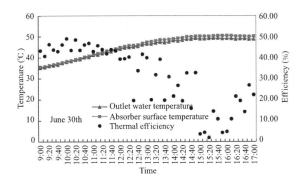

(b) The performance of the PV/T collector

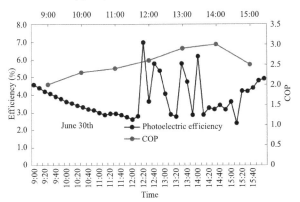

(c) The performance of the collector and ASHP

Figure 6-18　The results of a typical day (June 30th)

The system performance results on a sunny day (October 30th, 2018 in the heating season) are showed in Figure 6-19. On this day, heating, electricity and hot water were supplied. The average office room temperature was 19°C. As shown in Figure 6-19 (a), on October 30th, the ambient temperature had an upward trend from 14°C to the highest 23°C at 14:30, then had a slight decrease to nearly 20°C around 17:00. Solar irradiation was a typical trend in this day, and it initially increases then decreases. The highest irradiation value was 1100 W/m^2 at noon. From Figure 6-19 (b), it is seen that the absorber surface temperature and the outlet water temperature have a similar upward trend versus the ambient temperature change. Firstly, the absorber surface temperature increases from 15°C to 36°C from 9:00 to 15:20, and then it slightly declines to 35°C. The outlet water varies from 16 °C to the highest 33 °C, and finally down to 32 °C. The thermal efficiency is decreased with a small fluctuation from 65% to 10%, where the average is 35%. The reason for the decreasing thermal efficiency is that the solar irradiation was significantly decreased and the absorbed heat also decreased that afternoon. Therefore, the efficiency was reduced. The photoelectric efficiency and COP of ASHP change are presented in Figure 6-19 (c). As it is seen, the photoelectric efficiency varies with solar irradiation. The average photoelectric efficiency is 4% and the COP of ASHP is over 3, with an average of 3.3.

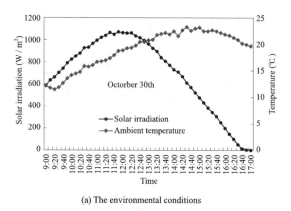

(a) The environmental conditions

Figure 6-19   The results of a typical day (October 30th) (Ⅰ)

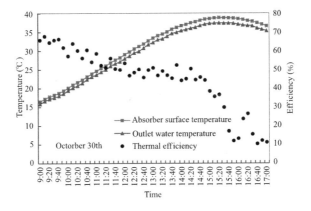

(b) The performance of the PV/T collector

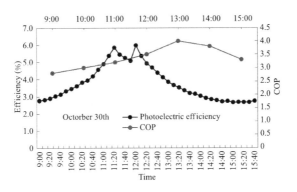

(c) The performance of the collector and ASHP

Figure 6-19 The results of a typical day (October 30th) (II)

For the extreme conditions, we also analyze the results on December 31th, 2018, where both heating and electricity were supplied. The average value of the office room temperature was 18°C. The external environmental conditions are presented in Figure 6-20 (a), and the ambient temperature is increased from −8°C to 0°C, and the average value of the ambient temperature is −4.5°C. As it is seen that the solar irradiation presents a similar trend to a normal day, it is increased first and then decreased. The peak value of solar irradiation is 900 W/m² at noon. Figure 6-20 (b) shows that the thermal efficiency fluctuates around 30%. There is a bigger temperature difference between the absorber surface and the outlet water temperature during this day compared with other days discussed above. Although the trends are the same, the values are lower than the abovementioned days, as the absorber surface temperature is from 13.5°C to 25°C and the outlet water temperature is 10°C to 17°C. The temperature difference between the ambient temperature and the outlet water temperature is also higher. Therefore, the COP of ASHP is on average of 3.7 and higher than before [in Figure 6-20 (c)]. Meanwhile, the photoelectric efficiency is also higher than other days, on average of 8% on December 31th, 2018.

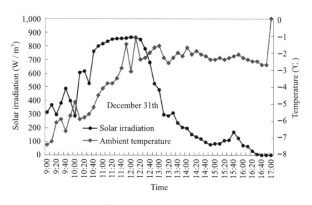

(a) The environmental conditions

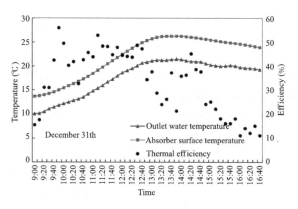

(b) The performance of PV/T collector

Figure 6-20 The results of a typical day (December 31st) (I)

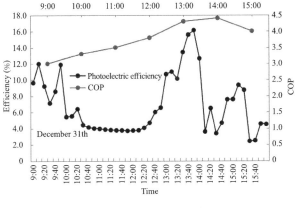

(c) The collector performance and ASHP

Figure 6-20　The results of a typical day (December 31st) (Ⅱ)

From the above results, we can see that the performance of the solar collector and the whole system including the COP, thermal efficiency, photoelectric efficiency, and so on in different environmental conditions varies significantly. Ambient temperature and solar irradiation changes cause a big influence on the performance of the whole system. The larger the difference between the ambient temperature and the outlet water temperature of the solar collector, the higher the COP of ASHP and electric production efficiency.

6.4.5　Economic analysis

The designed PV/T-ASHP system supplies heating, cooling, and hot water. For the office buildings, an air conditioner is generally used. This PV/T-ASHP system can be used both in new and renovated buildings. The total cost of the whole system installation is 8,000 CNY, and PV/T system electricity part costs 17625 CNY including the measurement equipment and weather station 2,600 CNY, and the heating part cost 10,000 CNY. The cost of ASHP system is 3,750 CNY.

In this case, an office room with a dimension of 37.5 m^2 is considered and the initial investment for the whole system is 850 CNY/m^2, while the installation cost (including equipment maintenance and labor fees) accounts for 10%. For renovated buildings, the initial investment except the ASHP system is 750 CNY/m² which includes the installation and maintenance cost.

To calculate the reduced running costs, electric price is considered. The total useful electricity is 6,819.5 kWh. In Beijing, the electric price is 0.4883 CNY/kWh for normal residential buildings. Therefore, for one whole year, the reduced electricity cost is 3,330 CNY (without considering the interest rate growth), and because the area for this room is 37.5 m², so the reduced running cost is 89 CNY/m². Note that the hot water supply to the office room is only used for washing hands, hence it is not concluded in the energy-saving part.

Based on the values of initial investment and reduced running costs for the whole year, the static payback time of this case is 9.5 years. Adding ASHP system, for renovation, the payback time becomes 8.5 years.

6.4.6 Environmental analysis

According to statistical analysis, in 2018, China's thermal power generation accounted for 73.32% of the total power generation, followed by hydropower generation, 16.24%, nuclear power generation, 4.33%, and wind power generation, 1.32%. Coal is the main energy source for thermal power generation. In this paper, we analyze the system's environmental benefits based on the standard coal, and each kilowatt-hour of electricity is equivalent to 0.4 kg standard coal saving.

The coefficient of energy consumption and pollution emissions is shown in Table 6-6. It is shown that each kilowatt-hour of electricity saving leads to 0.997 kg of carbon dioxide (CO_2) deduction, 0.03 kg of sulfur dioxide (SO_2) deduction, and 0.0015 kg of nitrogen oxides (NO_X) deduction. Note that 1 kg standard coal saving is equivalent to reducing 2.493 kg of carbon dioxide emission, 0.075 kg of sulfur dioxide emission, and 0.0375 kg of nitrogen oxides emission. In 2018, PV/T-ASHP system produced useful electricity 6819.5 kWh in total. According to calculation, it can save 2.73 t of standard coal. As a result, it can lead to 6.8 t of carbon dioxide deduction, 204.59 kg of sulfur dioxide deduction, and 102.29 kg of nitrogen oxides deduction throughout the year. Therefore, this system could effectively reduce environmental pollution.

Table 6-6 The coefficient of energy consumption and pollution emissions

Energy source	Carbon dioxide (CO_2)	Sulfur dioxide (SO_2)	Nitrogen oxides (NO_X)
1 kWh	0.997	0.03	0.015
1 kg standard coal	2.493	0.075	0.0375

6.5 Comparison of the proposed system with the previous research results

As shown in the relative published references, Alejandro et al. [26] conducted both simulations in TRNSYS and experimental studies for industrial use in Spain. The TRNSYS simulation results claimed that the system could achieve around 40% of electricity self-sufficiency. Compared to Alejandro et al. [26], this PV/T-ASHP system achieves higher (over 50%) electricity self-sufficiency. For the same system and in similar use, Wang et al. [19] and Raghed et al. [27] conducted the study on PV/T-ASHP system both in a cold region, Beijing and Toronto, respectively. In the heating seasons, the average COP of ASHP system was 3.03 in Wang et al. [19] Raghed et al. [27] also found the COP varied from 2.47 to 3.45. In this study, for the only heating season, the average COP of this system from October to February is up to 4, much higher than the above mentioned researches. The reason is mainly due to the design of the absorber, as the absorber in this paper is with pin-fins, which enhances the heat transfer of the absorber causing performance improvement. In general, this system performs well with additional characteristics, e.g., energy-saving, and environment friendly.

6.6 Summary

In this chapter, PV/T-ASHP system is presented based on existing studies. The system design and main results of these papers are given. The system performance common evaluation indicators mainly including COP are compared.

Moreover, a PV/T-ASHP system used for an office room is designed and set up. The performance of this system is presented including the COP, economic and environment influence. At last, its performance is also compared with existing PV/T-ASHP systems.

References

[1] He W, Zhang Y, Ji J. Comparative experiment study on photovoltaic and thermal solar system under natural circulation of water[J]. Applied Thermal Engineering, 2011, 31(16): 3369-3376.

[2] Charalambous P G, Maidment G G, Kalogirou S A, et al. Photovoltaic thermal (PV/T) collectors: A review[J]. Applied Thermal Engineering, 2007, 27(2-3): 275-286.

[3] Buker M S, Mempouo B, Riffat S B. Performance evaluation and techno-economic analysis of a novel building integrated PV/T roof collector: An experimental validation[J]. Energy and Buildings, 2014, 76: 164-175.

[4] Hassani S, Taylor R A, Mekhilef S, et al. A cascade nanofluid-based PV/T system with optimized optical and thermal properties[J]. Energy, 2016, 112: 963-975.

[5] Zhang X, Zhao X, Smith S, et al. Review of R&D progress and practical application of the solar photovoltaic/thermal (PV/T) technologies[J]. Renewable and Sustainable Energy Reviews, 2012, 16(1): 599-617.

[6] Kroiß A, Präbst A, Hamberger S, et al. Development of a seawater-proof hybrid photovoltaic/thermal (PV/T) solar collector[J]. Energy Procedia, 2014, 52: 93-103.

[7] Farshchimonfared M, Bilbao J I, Sproul A B. Full optimisation and sensitivity analysis of a photovoltaic–thermal (PV/T) air system linked to a typical residential building[J]. Solar Energy, 2016, 136: 15-22.

[8] Ibrahim A, Fudholi A, Sopian K, et al. Efficiencies and improvement potential of building integrated photovoltaic thermal (BIPVT) system[J]. Energy Conversion and Management, 2014, 77: 527-534.

[9] Foto Therm. 2016. Homepage [EB/OL]. Available: https://www.fototherm.com/en/ [2024-04-17].

[10] Hou L, Quan Z, Zhao Y, et al. An experimental and simulative study on a novel photovoltaic-thermal collector with micro heat pipe array (MHPA-PV/T)[J]. Energy and Buildings, 2016, 124: 60-69.

[11] Zhou J, Zhao X, Ma X, et al. Experimental investigation of a solar driven direct-expansion heat pump system employing the novel PV/micro-channels-evaporator modules[J]. Applied Energy, 2016, 178: 484-495.

[12] Jouhara H, Szulgowska-Zgrzywa M, Sayegh M A, et al. The performance of a heat pipe based solar PV/T roof collector and its potential contribution in district heating applications[J]. Energy, 2017, 136: 117-125.

[13] Chandrasekar M, Senthilkumar T. Experimental demonstration of enhanced solar energy utilization in flat PV (photovoltaic) modules cooled by heat spreaders in conjunction with cotton wick structures[J]. Energy, 2015, 90: 1401-1410.

[14] Xu P, Zhang X, Shen J, et al. Parallel experimental study of a novel super-thin thermal absorber based photovoltaic/thermal (PV/T) system against conventional photovoltaic (PV) system[J]. Energy Reports, 2015, 1: 30-35.

[15] Shen J, Zhang X, Yang T, et al. Characteristic study of a novel compact Solar Thermal Facade (STF) with internally extruded pin–fin flow channel for building integration[J]. Applied Energy, 2016, 168: 48-64.

[16] Kalogirou S A. Solar energy engineering: processes and systems[M]. Elsevier, 2009.

[17] Mathioulakis E, Panaras G, Belessiotis V. Uncertainty in estimating the performance of solar thermal systems[J]. Solar Energy, 2012, 86(11): 3450-3459.

[18] Çengel, Y.A. & Ghajar, A.J. 2015. Heat and Mass Transfer: Fundamentals and Applications [EB/OL]. Available: https://highered.mheducation.com/sites/9814595276 [2024-04-17].

[19] Wang G, Quan Z, Zhao Y, et al. Experimental study of a novel PV/T-air composite heat pump hot water system[J]. Energy Procedia, 2015, 70: 537-543.

[20] Rauschenbach H S. Solar cell array design handbook: the principles and technology of photovoltaic energy conversion[M]. Springer Science & Business Media, 2012.

[21] Qu M, Chen J, Nie L, et al. Experimental study on the operating characteristics of a novel photovoltaic/thermal integrated dual-source heat pump water heating system[J]. Applied Thermal Engineering, 2016, 94: 819-826.

[22] Besagni G, Croci L, Nesa R, et al. Field study of a novel solar-assisted dual-source multifunctional heat pump[J]. Renewable Energy, 2019, 132: 1185-1215.

[23] Li H, Sun Y. Operational performance study on a photovoltaic loop heat pipe/solar assisted heat pump water heating system[J]. Energy and Buildings, 2018, 158: 861-872.

[24] Lu S, Liang R, Zhang J, et al. Performance improvement of solar photovoltaic/thermal heat pump system in winter by employing vapor injection cycle[J]. Applied Thermal Engineering, 2019, 155: 135-146.

[25] Çengel, Y.A. & Ghajar, A.J. 2015. Heat and Mass Transfer: Fundamentals and Applications [EB/OL]. Available: https://highered.mheducation.com/sites/9814595276 [2024-04-17].

[26] Del Amo A, Martínez-Gracia A, Bayod-Rújula A A, et al. Performance analysis and experimental validation of a solar-assisted heat pump fed by photovoltaic-thermal collectors[J]. Energy, 2019, 169: 1214-1223.

[27] Kamel R, Ekrami N, Dash P, et al. BIPV/T+ ASHP: technologies for NZEBs[J]. Energy Procedia, 2015, 78: 424-429.

[28] Li H, Cao C, Feng G, et al. A BIPV/T system design based on simulation and its application in integrated heating system[J]. Procedia Engineering, 2015, 121: 1590-1596.

[29] Cao C, Li H, Feng G, et al. Research on PV/T–air source heat pump integrated heating system in severe cold region[J]. Procedia Engineering, 2016, 146: 410-414.

Chapter 7
ANN Prediction Study

7.1 Introduction

Photovoltaic/thermal (PV/T) system plays an important role in solar system development and application on reducing pollution and meeting the rapidly increasing energy demand. Modeling the system performance is required for the optimization and design of such a system. Its performance is affected by many factors, and therefore difficult to be estimated using the normal regression method. The requirement cost for investigating the performance of such systems is very high both in terms of time and monetary investments. Nevertheless, the comparisons among different systems with the same conditions are necessary for users and system performance prediction is therefore required. Artificial neural network (ANN) as an intelligent method is presented and conducted in this chapter. Environment factors including solar radiation, humidity, ambient temperature, wind speed, and wind direction and operation parameters of inlet/outlet water temperature, the collector surface temperature, and water volume are considered as the inputs for an office building in Beijing based on one year's experiment data. The electric and thermal efficiency of PV/T systems are as the outputs. The different numbers of factors and different ANN network constructions are conducted and compared.

7.1.1 Background

Along the rapid population growth, the energy demand is growing worldwide.[1] To deal with both of the large energy demand and environment problems, renewable energy resources have gained significant attentions in recent years, with the sustainable and non-pollution advantages.[2] Among all renewable energy resources, solar energy have attracted the greatest attention, as it is abundant and essentially inexhaustible. [3-5] There have been many different devices using solar energy, such as solar collectors, solar assisted heat pumps, PV/T system and so on. Among them, PV/T system, which absorbs the solar energy and produces both thermal energy and electricity, leads to better system performance. [6] PV/T system has been studied by using experimental and theoretical methods to figure out how the different parameters affect the performance of PV/T. [7-11] However, there are many limitations for experiments, such as its high cost and the general short experimental time , and the results are only suitable for limited similar system. Theoretical results are generally simplified models with some assumptions. [3] Facing this problem, various modeling approaches are proposed to predict the performance of PV/T system.

Multiple regression has been used in predicting the performance of the PV/T collector. For linear regression analysis, there are many researches focused on the influence of external environment changing or system design and operation on the performance of PV/T collector. [12-14] Helmers and Kramer [15] applied multi-linear regression model which was used and validated by measurement data. The input measured variables were the relevant ambient conditions, including the direct normal irradiation, local wind speed and ambient temperature. The outputs were the electrical and thermal power of PV/T collector. As a result, the normalized root mean square errors (NRMSE) were 1.9% and 2.9% for electric and thermal power, respectively.

as an intelligent based behavior of human brain technique, ANN has been verified to be a useful tool to model PV/T system especially solving the high nonlinear partial problem. Ability to find the relationship between inputs/outputs and also to have high speed simulations is the known benefits of these networks. [3,16-17] There have been many researches using ANN to model the PV/T both of electrical and thermal performance even under dynamic behavior. Kalani et al. [18] introduced a new method to model a photovoltaic thermal nanofluid based collector system (PVT/N) located at Ferdowsi university of Mashhad, Iran by using three different methods, namely, radial-basis function artificial neural network (RBFANN), multi-layer perception artificial neural network (MLPANN) and adaptive neuro fuzzy inference system (ANFIS) model. They identified the relationships between input and output parameters, including solar irradiation and ambient temperature for input parameters and fluid outlet temperature of the collector and electrical efficiency of PV for output parameters. RBFANN yielded significantly lower root mean square errors (RMSE) compared to that of the ANFiS ($p < 0.0003$). ANFiS was found to result in a higher performance than MLPANN ($p < 0.002$) and RBFANN ($p < 0.001$) in predicting fluid outlet temperature. Ghaini et al. [19] used an ANN to approximate the photovoltaic yield. The input parameters were the arrays of aspects ratio and mass flow rate. The output parameters were the approximate PV yield under four fluid flow configurations. Graditi et al. [20] compared two approaches: a physical modelization and ANN. A new hybrid method, hybrid physical artificial neural network (HPANN) was proposed and compared with multi-layer perceptron ANN (MLPANN). The inputs were ambient temperature, solar irradiation and clear shy solar irradiation model and outputs were the power production. The values of RMSE was less 10%. Gunasekar et al. [21] developed an ANN to predict the energy performance of a PV/T evaporator in solar assisted heat pump system. The input variables are solar intensity, ambient temperature, relative humidity and wind velocity, each of them has three values. The output variables are evaporator heat gain, solar energy input ratio, panel efficiency and panel temperature. The prediction from the ANN illustrate that photovoltaic panel temperature yields maximum R^2 value of 0.9999, minimum RMSE value of 0.014 °C and minimum COV value of 0.0545. Kamthania and Tiwari [22] used ANN to analyze the performance of a semi transparent hybrid PV/T double pass air collector under four weather conditions at New Delhi. The ambient air temperature, global solar radiation, diffuse radiation and number of clear days are regarded as the input parameters. Electrical energy, thermal energy, overall thermal energy and exergy are the estimated parameters. The RMSE varies from 0.10% to 2.23%. Ravaee et al. [23] used ANN to develop both thermal and electrical modeling of PV/T collector. The inputs are ambient temperature of collector, cell temperature, fluid temperature at duct inlet, fluid velocity in duct, solar identity and time. The output are the thermal efficiency and electrical efficiency. R^2 is 0.9976, with RMSE and COV values of 0.3974 and 0.0526. Ammar et al. [24] used ANN based control to optimize the thermal and electric power of a PV/T panel for a given solar radiation and ambient temperature. Input parameters include the glazing temperature, solar cell, absorber plate and water circulation. The outputs are electric power and thermal profit. NMBE is −13.05%. Khatib et al. [25] used two different ANNs generalized regression artificial neural network and cascaded forward neural network to predict current-voltage (I-V) curve, while Karatepe et al [26], Celik [27] and Bonanno et al. [28] also used ANN to conduct the I-V curve extracting of PV solar cells. The input variables are solar radiation, ambient temperature and open circuit voltage and short circuit current. As a result, the average mean absolute percentage error, mean bias error and root mean square error are 1.09%, 0.0229 A and 0.0336 A, respectively.

Aside from above mentioned methods, genetic algorithm (GA) was proposed by Singh et al. [29-30] to improve the efficiency of PV/T system in India. The influencing parameters are length and depth of the channel, velocity of air fluid flowing into the channel, thickness of the tedlar and glass, temperature of inlet fluid. The objective functions are the overall exergy efficiency and overall thermal efficiency. The data was obtained for whole one year, while the used data were the average parameters obtained at 11:00 AM for each month. The overall thermal efficiency and overall exergy efficiency improvements were 13.14% and 4.6%, respectively. Singh et al. [31] also compared the GA and GA-FS (generic algorithm-fuzzy system).It has been observed that GA-FS approach outperforms the standard GA (Genetic Algorithm) by converging more quickly due to the integration of a fuzzy knowledge base, which also results in less time required for the identification of optimized system parameters.Kim et al. [32] applied back-propagation network (BPN) model incorporating genetic algorithms (GAs) to cost estimation. Sobhnamayan et al. [33] used GA to optimize the exergy efficiency of PV/T water collector.

Evolutionary algorithm (EA) was also used to optimize the single channel glazed PV/T array by Singh et al. [34] The objective function is the overall exergy gain. Seven design parameters are the influence factors including length of the channel, mass flow rate of flowing fluid, velocity of flowing fluid, convective heat transfer coefficient through the tedlar, overall heat transfer coefficient between solar cell to ambient through glass cover, overall back loss heat transfer coefficient from flowing fluid to ambient and convective heat transfer coefficient of tedlar.

A summary of the main results of aforementioned is presented in Table 7-1. After reviewing the previous papers, a few key points are identified as below:

• Most of the employed methods used to optimize or predict the performance of PV/T are ANN and MLR. Although there are many researches on the analysis of PV/T system [10], the researches use different algorithm based on ANN, but it still requires further exploration. [17] As we can see that there are insufficient datasets in previous researches, even the datasets in some researches are only few days measurement.[3, 21, 28] So the application of ANN still needs further investigation based on large data.

• Different algorithm methods and indicators have been adopted by researchers, but most of them are only be applied or validated by one model construction. However, the number of influence factors are stable for each research. There are no researches focusing on the comparison between the effects of different influence factors combination on the results.

• There have been many researches of ANN on solar system. [3, 35] However, the researches on the performances of PV/T systems is still inadequate in China.

This chapter aims to study the performance of PV/T collector using an ANN algorithm with a back propagation (BP) algorithm based on a PV/T system with whole-year experimental measurement in Beijing. Moreover, the results are compared with different number factors' combination to find the most accurate model.

Chapter 7
ANN Prediction Study

Table 7-1 Overviewed of main results of previous researches of PV/T collector models

Reference	Model	Location	Data	Input parameters	Output parameters	Accuracy
Helmers and Kramer	MLR	Fraunhofer, Germany	15,294	direct normal irradiation, local wind speed and ambient temperature	electrical and thermal power	NRMSE=1.9% (electric), 2.9% (thermal power)
Singh et al.	GA	New Delhi, India	one day	length and depth of the channel, velocity of air fluid flowing into the channel, thickness of the tedlar and glass, temperature of inlet fluid	overall exergy efficiency and overall thermal efficiency	overall thermal efficiency and overall exergy efficiency improvements were 13.14% and 4.6%
Kim et al.	BPN-GA	Seoul, Korea.	530 buildings	building size, location, number of storeys	construction costs	MAER=3.16
Singh et al.	EA	New Delhi, India		seven design variables	overall exergy gain	
Kalani et al.	RBFANN; MLPANN; ANFIS	Iran	10 days with each day 13 points	Solar irradiation, ambient temperature	fluid outlet temperature; electrical efficiency	R^2>0.98; RMSE <0.4%
Ghaini et al.	ANN		288 simulation data	arrays of aspects ratio, mass flow rate	approximate PV yield under four fluid flow configurations	r=0.984
Graditi et al.	a physical modelization and HPANN	ENEA Portici Research Center	500 epochs	ambient temperature, solar irradiation and clear shy solar irradiation	power production	RMSE <10%

Continued

Reference	Model	Location	Data	Input parameters	Output parameters	Accuracy
Gunasekar et al.	ANN	Coimbatore city in India	1 day	solar intensity, ambient temperature, relative humidity and wind velocity	evaporator heat gain, solar energy input ratio, panel efficiency and panel temperature	max $R2$ 0.9999, min RMS=0.014 °C and COV=0.0545
Kamthania and Tiwari	ANN	New Delhi	2,000	ambient air temperature, global solar radiation, diffuse radiation and number of clear days	electrical energy, thermal energy, overall thermal energy and exergy	RMSE 0.10%–2.23%
Ravaee et al.	ANN	Zahedan		ambient temperature of collector, cell temperature, fluid temperature at duct inlet, fluid velocity in duct, solar identity and time	thermal efficiency and electrical efficiency	R^2 = 0.9976, RMSE =0.3974 COV= 0.0526
Ammar et al.	ANN	Tunisia	100	glazing temperature, solar cell, absorber plate and water circulation	electric power and thermal profit	NMBE= −13.05%
Khatib et al.	generalized regression ANN and cascaded forward NN	Nablus, Palestine	12,100	solar radiation, ambient temperature and open circuit voltage and short circuit current	I-V curve	MAPE= 1.09%, MBE=0.0229 A, RMSE=0.0336 A

Chapter 7
ANN Prediction Study

7.1.2 Data collecting

A PV/T system is mainly consisted of PV/T collector, water tank and water circulation heat pump system is measured for the whole year of 2019. The schematic of the system is shown in Figure 7-1. The inclined angle of the collector is 47°, facing south. The total surface area for the PV/T collector with two arrays is 2.6 m^2, and that for each one is 1.3 m^2. The water tank volume is 100 L, which is used for hot water storage. The refrigerant used in the ASHP system is $R22$. The experiment is located in an office room near Beijing (117°E, 40°N), a cold region of China.

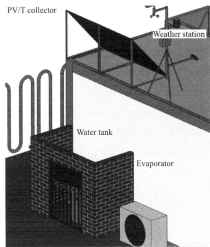

Figure 7-1 The picture of part of PV/T-ASHP system

The external environment conditions are measured by the weather station installed nearby the PV/T collector (Figure 7-1). The measurements include solar radiation (W/m^2), external humidity (%), ambient temperature (°C), the wind speed (m/s) and wind direction (°) for environment parameters and plate surface temperature (°C), the inlet water volumetric flow (kg/h) and inlet/outlet water temperature (°C) of solar collector. We use temperature probe with the accuracy of ± 0.01°C for the measurement. The water flow used ultrasonic heat meter, with the range 0.07-7 m³/h and the accuracy of level-2, ±0.14 m³/h. Experiments were operated during 9:00 to 17:00 in the whole year of 2018, and time interval of all measurements was 10 min.

7.1.3 Methodology

Artificial neural network (ANN)

ANN is a processor which is widely parallel distributed and is composed of simple processing units (called neurons). [36] It is popularly used to solve complex functions in various applications such as high-speed information proceeding, mapping capabilities, forecasting, optimization and so on to reduce the cost and time. [37] There are many different ANN types, including multi-layer perceptron ANN (MLPANN), wavelet neural network (WNN), radial basic function ANN (RBFANN), and Elman neural network (ENN) and so on. [3] The common used types are MLP and RBF, and MLP has been used successfully to solar energy system. [3, 18-19, 38-39] BP neural network is widely used with ANN, with two process: the forward signal transmission and reverse estimated error transmission. The structure of ANN with BP algorithm consists input layer, one or more hidden layers and output layer, as shown in Figure 7-2.

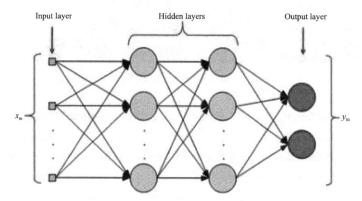

Figure 7-2 Graph of multi-layer artificial neural network based on large data [40]

To predict the performance of PV/T collector, the collected data is split into three subset: training, validation and testing with the ratio 3:1:1. And the total data size is 14,400, totally 300 days in 2018. Majority of data are used for training to test the network, validation data is used to mitigate overtraining of the network to gain acceptable validation error. Then the testing data is used in trained network. The prediction output value is compared with the collected data to find the best and most suitable ANN network, different structure of ANN with one hidden layer and two hidden layers, which are commonly used types in existing researches [3], are set and the accuracy are compared.

In this study, the measurement data consists of environment factors and operation parameters. Environmental factors are solar radiation (θ_{sr}), outdoor humidity (θ_{eh}), ambient temperature (θ_{at}), wind direction (θ_{wd}) and wind speed (θ_{ws}), and operation system parameters contains water volumetric flow (θ_{wf}), outlet water temperature (θ_{wt}), the plate surface temperature (θ_{pt}), and inlet water temperature (θ_{it}). To verify the effect of the environmental and operational factors on efficiency, we divide these factors into three groups. The first contains all factors (totally nine), the second is only the operation factors (five factors) and only environmental factors (four factors).

7.1.4 Evaluation criteria

The estimation indicator on how well the equation fits the data is expressed by the coefficient of determination, R^2 as shown in Equation (7-1). It varies from 0 to 1, higher the value (more near 1), better the fits. And the difference between the observed and fitted values known as the root mean squared error (MSE) is calculated to determine the accuracy of the prediction, the equations are as below:

$$R^2 = \frac{[\sum_{1}^{n_{sample}} (y_{observed} - \overline{y_{observed}})(y_{prediction} - \overline{y_{prediction}})]^2}{\sum_{1}^{n_{sample}} (y_{observed} - \overline{y_{observed}})^2 \times \sum_{1}^{n_{sample}} (y_{prediction} - \overline{y_{prediction}})^2} \quad (7\text{-}1)$$

$$\text{RMSE} = \sqrt{\frac{1}{n_{\text{sample}}} \sum_{1}^{n_{\text{sample}}} (y_{\text{observed}} - \overline{y_{\text{prediction}}})^2} \qquad (7\text{-}2)$$

Where, n_{sample} is the number of samples, y_{observed} is the true observed/collected data of samples and $\overline{y_{\text{observed}}}$ is the average value of true data, $y_{\text{prediction}}$ is the predicted/estimated value and $\overline{y_{\text{prediction}}}$ is the average value of prediction.

Besides the R^2 and MSE, the root mean absolute error (MAE) is also calculated for ANN with BP algorithm, as shown in Equation (7-3).

$$\text{MAE} = \frac{1}{n_{\text{sample}}} \sum_{1}^{n_{\text{sample}}} |y_{\text{observed}} - \overline{y_{\text{observed}}}| \qquad (7\text{-}3)$$

7.2 Data observation

In order to analyze the real-time performance of this PV/T-ASHP system, a set of experimental data under typical climate conditions in September 30th, 2019 are selected. The results are shown below.

During the heating seasons, the results on Sunday (September 30th, 2019) are shown in Figure 7-3. As shown in Figure 7-3 (a), in September 30th, 2019, ambient temperature had an upward trend from 10 °C to the highest 20 °C at 16:00, then had a slight decrease to around 18 °C at the end time 17:00. Solar irradiation first increased then decreased. The highest was 1,100 W/m² at 12:00. From Figure 7-3 (b), we can see that the surface temperature of the absorber and the outlet water temperature had a similar variation profile as the ambient temperature change. Outlet water temperature firstly increased from 15 °C to 33 °C during 9:00 to 15:20, then it was slightly down to 32 °C. The absorber surface temperature varied from 32 °C to the highest 74 °C, then to 72 °C. The thermal efficiency decreased with a small fluctuation from 65% to 10%, on average of 45%. The reason for decrease of thermal efficiency is that after 12:00, the solar irradiation decreased dramatically. It caused a decrease of absorbed heat, and hence the efficiency reduce. And other measurement data including humidity, wind speed and wind direction and so on is shown in Table 7-2.

Chapter 7
ANN Prediction Study

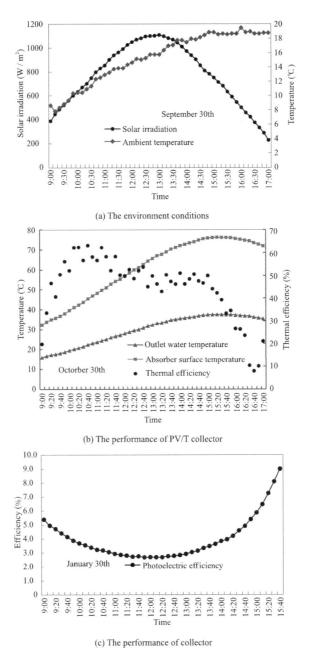

(a) The environment conditions

(b) The performance of PV/T collector

(c) The performance of collector

Figure 7-3 The results in one typical day

113

Table 7-2 The typical day measurement results

Time (min)	Inlet water temperature (°C)	Volumetric flow (kg/h)	Humidity (%)	Wind speed (m/s)	Wind direction (°)
9:00:00	15.94	1,509	51.6	0	164
9:10:00	16.58	1,483	55.9	0.5	196
9:20:00	17.15	1,511	57.1	0.5	330
9:30:00	17.63	1,473	53	0.3	263
9:40:00	18.05	1,489	43.5	0.2	148
9:50:00	18.64	1,490	45.6	0.1	145
10:00:00	19.36	1,493	42.7	0.2	154
10:10:00	20.04	1,490	42.6	0.1	12
10:20:00	20.74	1,477	37.6	0.4	159
10:30:00	21.32	1,498	30.8	0.2	349
10:40:00	22.07	1,521	32.4	0.3	320
10:50:00	22.94	1,496	32	0.3	349
11:00:00	23.72	1,494	27.9	0.6	176
11:10:00	24.4	1,497	30.7	0.4	143
11:20:00	25.12	1,508	30.3	0.4	151
11:30:00	25.82	1,471	29.6	0.4	347
11:40:00	26.6	1,480	29.5	0.3	344
11:50:00	27.22	1,507	26.9	0.4	164
12:00:00	28.04	1,510	27.2	0.4	28
12:10:00	28.79	1,518	26.9	0.3	110
12:20:00	29.53	1,504	27.3	0.5	32
12:30:00	30.27	1,507	26.8	0.4	157
12:40:00	30.86	1,482	24.4	0.2	340
12:50:00	31.61	1,534	25	0.4	89
13:00:00	32.35	1,522	25	0.5	83
13:10:00	32.92	1,491	25	0.4	138

Continued

Time (min)	Inlet water temperature (°C)	Volumetric flow (kg/h)	Humidity (%)	Wind speed (m/s)	Wind direction (°)
13:20:00	33.53	1,507	21.7	0.3	139
13:30:00	34.07	1,518	23.2	0.5	15
13:40:00	34.68	1,520	22.5	0.1	160
13:50:00	35.28	1,525	21.7	0.5	13
14:00:00	35.77	1,499	21.7	0.1	6
14:10:00	36.03	1,529	22.3	0.1	163
14:20:00	36.46	1,515	19.2	0.3	174
14:30:00	36.77	1,517	19.7	0.1	148
14:40:00	37.28	1,515	19.3	0.4	147
14:50:00	37.61	1,508	21.6	0.1	158
15:00:00	37.79	1,500	19.9	0.2	143
15:10:00	38.05	1,523	20.7	0.3	164
15:20:00	38.14	1,524	21.3	0.3	136
15:30:00	38.07	1,517	21.6	0.2	141

7.3 ANN

The ANN models developed in this Chapter can be divided into three groups, which have different number of input layers. The number of input layer factors are 4 [water volumetric flow (θ_{wf}), outlet water temperature (θ_{wt}), the plate surface temperature (θ_{pt}), inlet water temperature (θ_{it})], 5 [solar radiation (θ_{sr}), outdoor humidity (θ_{eh}), ambient temperature (θ_{at}), wind direction (θ_{wd}), wind speed (θ_{ws})] and 9 (all factors including environment and operation).

In this study, the number of the input layer neurons is different, whereas the output is two neurons, thermal efficiency and electric efficiency. Different hidden layer and neurons of each hidden layer can lead to different results, as it affects the speed of the convergence and generalization of BP network. In this book, the BP neural network is designed to have one hidden layer N-X-1 or two hidden layers N-X-Y-1. That is, the input layer has N

nodes, and one hidden layer has X nodes. If there is the second hidden layer, the second layers has Y nodes, and the output layers has 1 node. To gain the best accurate results, different BP network with one hidden layer for different input layers and two hidden layers with eight input layers are trained and validated. The number nodes of hidden layers are chose as 5, 10, 15, 20 and 25, respectively.

7.4 Results

7.4.1 ANN model with one hidden layer

The results in one hidden layer with different nodes are shown in Table 7-3 and Table 7-4 for electric and thermal efficiency, respectively. For accuracy, we take the average value from training, testing and validation process. It is clear that the nodes number almost has no influence for electric efficiency and thermal efficiency. But with the same nodes, the results vary a lot and the variation trend for both electric and thermal efficiency are different. The reason is that the influence of environment and operation parameters differs in the results of electrical and thermal efficiency. While the electric efficiency is significantly affected by environment factors, and thermal efficiency is more affected by operation factors than environment factors, through comparing the results vary the groups 4 and 5 in both Table 7-3 and Table 7-4.

Table 7-3　R^2, RMSE and RMAE with one hidden layer for electric efficiency

Nodes	R^2			RMSE (%)			RMAE		
	9	4	5	9	4	5	9	4	5
5	0.9747	0.9753	0.7121	0.9781	0.9577	2.5032	0.6022	0.5588	1.2982
10	0.9841	0.9692	0.7355	0.9363	0.9615	2.4376	0.5697	0.5762	1.2880
15	0.9716	0.9798	0.7259	0.9248	0.9514	2.4098	0.5416	0.5628	1.2639
20	0.9772	0.9818	0.7291	0.8981	0.9391	2.4603	0.5488	0.5747	1.2891
25	0.9716	0.9685	0.7362	0.9689	0.9332	2.3951	0.5821	0.5688	1.2695

Chapter 7
ANN Prediction Study

Table 7-4 R^2, RMSE and RMAE with one hidden layer for thermal efficiency

Nodes	R^2			RMSE (%)			RMAE		
	9	4	5	9	4	5	9	4	5
5	0.7743	0.4810	0.7467	9.8135	13.3691	10.6664	2.5182	3.1148	2.6701
10	0.7971	0.4505	0.7272	9.8139	13.8220	10.6001	2.5557	3.1518	2.6617
15	0.7992	0.4933	0.7288	9.4591	13.3579	10.6149	2.4957	3.1134	2.6745
20	0.8037	0.5019	0.7346	9.3773	13.3439	10.7131	2.4901	3.0961	2.6950
25	0.8025	0.5073	0.7279	9.2248	13.3485	10.7014	2.4932	3.1015	2.6824

Figure 7-4 and Figure 7-5 show comparison between the results of R^2, RMSE and RMAE. Figure 6-4 shows that, the variation of the node's number of one hidden layer caused slight influence on the prediction models, as the R^2, RMSE and RMAE only experience small variations in each group for electric efficiency. The best ANN model is the one with 9 input factors, then the four environment factors, and the last five operation factors. As the average RMSE values are less 1.0% for input groups 9 and 4, while it is nearly 2.5% for group 5. And the RMAE for groups 9 and 4 are nearly 0.55, while for group 5 is nearly 1.3. From these results, we can also see that the difference between groups 9 and 4 is small, leading to that the electric efficiency is greatly affected by environment factors.

It is clear that, in Figure 7-5, the performance for different groups is group 9 > group 5 > group 4. From Figure 7-5, we can also see that the thermal efficiency is affected stronger by operation factor than environment factors. Comparing with electric efficiency results, the accuracy for thermal efficiency is lower. The highest R^2 for thermal efficiency is nearly 0.8, the lowest RMSE percentage is nearly 8.6 and lower RMAE is nearly 2.5 with input group 9.

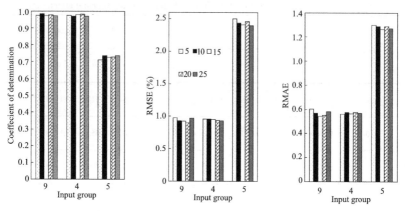

Figure 7-4　R^2, RMSE and RMAE for electric efficiency with one hidden layer

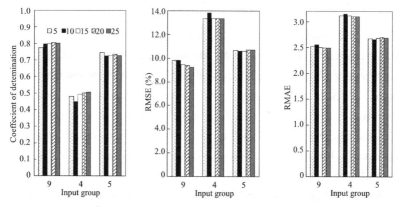

Figure 7-5　R^2, RMSE and RMAE for thermal efficiency with one hidden layer

7.4.2　ANN model with two hidden layers

Based on the results of R^2, RMSE and RMAE in Section 7.5.1, the more factors generate better performance, although the R^2, RMSE and RMAE difference between group 9 and group 4 is small for electric efficiency. Combination of two output parameters show the difference is large, especially for thermal efficiency. For prediction models of ANN with BP algorithm with two hidden layers, we take the input in group 9 to find the best network as example.

Most of network construction of 9-X-Y-1 with the X/Y vary from 5 to 25, but some of networks do not work well for prediction of the performance of electric and thermal efficiency in this PV/T system. The results can be seen from Table 7-5. This also shows that, not all networks are suitable for one system, it is necessary to optimize of the ANN network firstly before using for special system. And from Table 7-5, it is easily to find that the R^2 difference is minor for different nodes of hidden layers.

Table 7-5 R^2 for network with two hidden layers

	Electric efficiency					Thermal efficiency				
	X=5	X=10	X=15	X=20	X=25	X=5	X=10	X=15	X=20	X=25
Y=5	0.9669	0.9674	0.9624	0.9652	—	0.9628	0.9642	0.9689	—	—
Y=10	0.9656	0.9695	0.9704	0.0000	0.9754	—	0.9734	0.9646	0.9660	0.9665
Y=15	0.9749	0.9618	0.9649	0.9661	0.9627	0.9592	0.9670	0.9665	0.9686	0.9607
Y=20	0.9740	0.9624	0.9669	0.9697	0.9610	0.9704	0.9709	0.9746	0.9596	0.9647
Y=25	0.9736	—	0.9670	0.9652	0.9678	0.9702	0.9595	0.9648	0.9667	0.9652

The results of RMSE and RMAE with different hidden layer nodes are shown in graphic representation, identifying the output difference. The results of RMSE and RMAE for electric and thermal efficiency, respectively, are shown in Figure 7-6 and Figure 7-7. For electric efficiency (as shown in Figure 7-8), RMSE varies from 0.75% to 0.96%, lower than 1%, and the RMAE varies from 0.5 to 0.68, lower than 0.7 for all networks. The results show that the reality and prediction have a good agreement by using ANN network with two hidden layers. And the best network construction is the 9-25-25-1, as it has the lowest RMSE (0.76%). For thermal efficiency, the RMSE changes from 0.8% to 1.0%, and RMAE changes from 0.5 to 0.68. The beast network construction is 9-10-10-1, with the lowest value of RMSE (0.805%). And compared with electric efficiency prediction, the fluctuations of RMSE and RMAE are little larger for thermal efficiency.

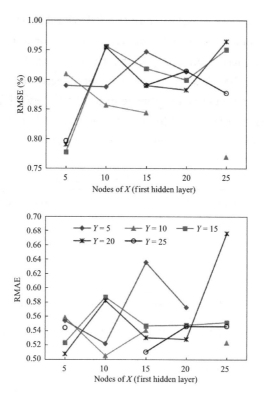

Figure 7-6   RMSE and RMAE of electric efficiency with two hidden layers

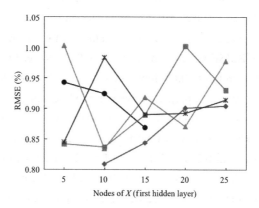

Figure 7-7   RMSE and RMAE of thermal efficiency with two hidden layers (I)

Chapter 7
ANN Prediction Study

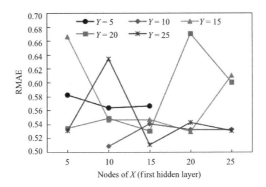

Figure 7-7 RMSE and RMAE of thermal efficiency with two hidden layers (II)

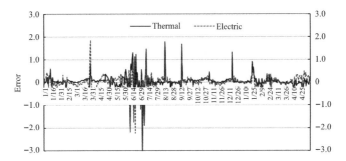

Figure 7-8 The error value of thermal and electric efficiency with 9-15-15-1 network

Figure 7-8 shows the error between prediction and measurement values (only 500 chosen in this figure) with 9-15-15-1 BP network in chosen days in 2018. The error between reality and prediction of electric and thermal efficiency varies from −3 to 2, on average of -0.87, which also means there is a good agreement between the reality and prediction.

7.5 Results comparison

Compared the results with nine input layers, it can be seen that the ANN network with two hidden layers performs better than one hidden layer, especially for thermal efficiency.

The best ANN construction for electric efficiency with one hidden layer is 9-20-1, with the lowest RMSE 0.898%. The 9-X-1 network for electric

121

efficiency shows that R^2 fluctuates around 0.97, and RMSE is lower than 0.98% and the RMAE is lower than 0.6. But for 9-X-Y-1 type, the results for electric shows that the average R^2 is about 0.97, but the RMSE is lower than 9-X-1 type, as the average value of RMSE is lower than 0.95%, most of them are only around 0.9%. And the RMAE is also lower, as most of them are about 0.55.

For thermal efficiency, the best ANN network with one hidden layer is 9-25-1, with the lower RMSE 9.225%. For 9-X-1 type for thermal efficiency, there is a trend that the more nodes of X, the better performance of the model. The R^2 is around 0.8, RMSE varies from 9.2% to 9.8%, and RMAE is around 2.5 for thermal efficiency of all 9-X-1 types. However, the R^2 for 9-X-Y-1 type are all over 0.9, most of them are around 0.97, and the RMSE are lower than 1.0%. RMAE are also much lower, as they are all smaller than 0.70 and most of them are near 0.55.

Compared with previous researches that the outputs are the similar, the results in this paper are better than previous researches.[3, 15, 20, 22] The NRMSE are 1.9% for electric and 2.9% for thermal power.[15] The RMSE is lower than 10% for power production [20] and 0.10%-2.23% for overall thermal energy and exergy.[22] The NMBE is -13.05% for.[3] There may be caused by few reasons. Firstly, this may be caused by the number of inputs, as inputs in this paper contain both environment and operation parameters, some of them [3, 15, 20, 22] only consider one or the number is smaller than this paper. This conclusion also dedicates that the more factors are considered, the better performance of the prediction model. Secondly, for previous researches, they only chose one model type without comparing the different constructions of ANN network. So there maybe exist improvement for their ANN model construction. This suggests that the researches need to find the best suitable ANN network before conducting, though the error may be not big.

7.6 Summary

In this chapter, the system's performance prediction method-ANN is presented. Firstly, the existing studies are given. And then based on case study system in chapter 5, the accuracy of ANN on PV/T-ASHP system are conducted.

References

[1] Şener Ş E C, Sharp J L, Anctil A. Factors impacting diverging paths of renewable energy: A review[J]. Renewable and Sustainable Energy Reviews, 2018, 81: 2335-2342.

[2] Guven G, Sulun Y. Pre-service teachers' knowledge and awareness about renewable energy[J]. Renewable and Sustainable Energy Reviews, 2017, 80: 663-668.

[3] Elsheikh A H, Sharshir S W, Abd Elaziz M, et al. Modeling of solar energy systems using artificial neural network: A comprehensive review[J]. Solar Energy, 2019, 180: 622-639.

[4] Chen Q F, Du S W, Yuan Z X, et al. Experimental study on performance change with time of solar adsorption refrigeration system[J]. Applied Thermal Engineering, 2018, 138: 386-393.

[5] Sharma A K, Sharma C, Mullick S C, et al. Solar industrial process heating: A review[J]. Renewable and Sustainable Energy Reviews, 2017, 78: 124-137.

[6] Sardarabadi M, Passandideh-Fard M, Heris S Z. Experimental investigation of the effects of silica/water nanofluid on PV/T (photovoltaic thermal units)[J]. Energy, 2014, 66: 264-272.

[7] Dubey S, Sandhu G S, Tiwari G N. Analytical expression for electrical efficiency of PV/T hybrid air collector[J]. Applied Energy, 2009, 86(5): 697-705.

[8] Atheaya D, Tiwari A, Tiwari G N, et al. Analytical characteristic equation for partially covered photovoltaic thermal (PVT) compound parabolic concentrator (CPC)[J]. Solar Energy, 2015, 111: 176-185.

[9] Yazdanpanahi J, Sarhaddi F, Adeli M M. Experimental investigation of exergy efficiency of a solar photovoltaic thermal (PVT) water collector based on exergy losses[J]. Solar Energy, 2015, 118: 197-208.

[10] Zhang X, Zhao X, Smith S, et al. Review of R&D progress and practical application of the solar photovoltaic/thermal (PV/T) technologies[J]. Renewable and Sustainable Energy Reviews, 2012, 16(1): 599-617.

[11] Guo C, Ji J, Sun W, et al. Numerical simulation and experimental validation of tri-functional photovoltaic/thermal solar collector[J]. Energy, 2015, 87: 470-480.

[12] Farshchimonfared M, Bilbao J I, Sproul A B. Channel depth, air mass flow rate and air distribution duct diameter optimization of photovoltaic thermal (PV/T) air collectors linked to residential buildings[J]. Renewable Energy, 2015, 76: 27-35.

[13] Chen Y, Athienitis A K, Galal K. Modeling, design and thermal performance of a BIPV/T system thermally coupled with a ventilated concrete slab in a low energy solar house: Part 1, BIPV/T system and house energy concept[J]. Solar Energy, 2010, 84(11): 1892-1907.

[14] Shen J, Zhang X, Yang T, et al. Characteristic study of a novel compact Solar Thermal Facade (STF) with internally extruded pin–fin flow channel for building integration[J]. Applied Energy, 2016, 168: 48-64.

[15] Helmers H, Kramer K. Multi-linear performance model for hybrid (C) PVT solar collectors[J]. Solar Energy, 2013, 92: 313-322.

[16] Papari M M, Yousefi F, Moghadasi J, et al. Modeling thermal conductivity augmentation of nanofluids using diffusion neural networks[J]. International Journal of Thermal Sciences, 2011, 50(1): 44-52.

[17] Mohanraj M, Jayaraj S, Muraleedharan C. Applications of artificial neural networks for thermal analysis of heat exchangers–a review[J]. International Journal of Thermal Sciences, 2015, 90: 150-172.

[18] Kalani H, Sardarabadi M, Passandideh-Fard M. Using artificial neural network models and particle swarm optimization for manner prediction of a photovoltaic thermal nanofluid based collector[J]. Applied Thermal Engineering, 2017, 113: 1170-1177.

[19] Ghani F, Duke M, Carson J K. Estimation of photovoltaic conversion efficiency of a building integrated photovoltaic/thermal (BIPV/T) collector array using an artificial neural network[J]. Solar Energy, 2012, 86(11): 3378-3387.

[20] Graditi G, Ferlito S, Adinolfi G, et al. Energy yield estimation of thin-film photovoltaic plants by using physical approach and artificial neural networks[J]. Solar Energy, 2016, 130: 232-243.

[21] Gunasekar N, Mohanraj M, Velmurugan V. Artificial neural network modeling of a photovoltaic-thermal evaporator of solar assisted heat pumps[J]. Energy, 2015, 93: 908-922.

[22] Kamthania D, Tiwari G N. Performance analysis of a hybrid photovoltaic thermal double pass air collector using ANN[J]. Applied Solar Energy, 2012, 48: 186-192.

[23] Ravaee H, Farahat S, Sarhaddi F. Artificial neural network based model of photovoltaic thermal (pv/t) collector[J]. The Journal of Mathematics and Computer Science, 2012, 4(3): 411-417.

[24] Ammar M B, Chaabene M, Chtourou Z. Artificial neural network based control for PV/T panel to track optimum thermal and electrical power[J]. Energy Conversion and Management, 2013, 65: 372-380.

[25] Khatib T, Ghareeb A, Tamimi M, et al. A new offline method for extracting IV characteristic curve for photovoltaic modules using artificial neural networks[J]. Solar Energy, 2018, 173: 462-469.

[26] Karatepe E, Boztepe M, Colak M. Neural network based solar cell model[J]. Energy Conversion and Management, 2006, 47(9-10): 1159-1178.

[27] Celik A N, Acikgoz N. Modelling and experimental verification of the operating current of mono-crystalline photovoltaic modules using four-and five-parameter models[J]. Applied Energy, 2007, 84(1): 1-15.

[28] Bonanno F, Capizzi G, Graditi G, et al. A radial basis function neural network based approach for the electrical characteristics estimation of a photovoltaic module[J]. Applied Energy, 2012, 97: 956-961.

[29] Singh S, Agarwal S, Tiwari G N, et al. Application of genetic algorithm with multi-objective function to improve the efficiency of glazed photovoltaic thermal system for New Delhi (India) climatic condition[J]. Solar Energy, 2015, 117: 153-166.

[30] Singh S, Agrawal S, Tiwari A, et al. Modeling and parameter optimization of hybrid single channel photovoltaic thermal module using genetic algorithms[J]. Solar Energy, 2015, 113: 78-87.

[31] Singh S, Agrawal S. Parameter identification of the glazed photovoltaic thermal system using Genetic Algorithm–Fuzzy System (GA–FS) approach and its comparative study[J]. Energy Conversion and Management, 2015, 105:

763-771.

[32] Gwang H K, Jie E Y, Sung H A. Neural network model incorporating a genetic algorithm in estimating construction costs[J]. Building and Environment, 2004, 39(11): 1333-1340.

[33] Sobhnamayan F, Sarhaddi F, Alavi M A, et al. Optimization of a solar photovoltaic thermal (PV/T) water collector based on exergy concept[J]. Renewable Energy, 2014, 68: 356-365.

[34] Singh S, Agrawal S, Gadh R. Optimization of single channel glazed photovoltaic thermal (PVT) array using Evolutionary Algorithm (EA) and carbon credit earned by the optimized array[J]. Energy Conversion and Management, 2015, 105: 303-312.

[35] Antonanzas J, Osorio N, Escobar R, et al. Review of photovoltaic power forecasting[J]. Solar Energy, 2016, 136: 78-111.

[36] Haykin S. Neural networks and learning machines, 3/E[M]. Pearson Education India, 2009.

[37] Jani D B, Mishra M, Sahoo P K. Application of artificial neural network for predicting performance of solid desiccant cooling systems–A review[J]. Renewable and Sustainable Energy Reviews, 2017, 80: 352-366.

[38] Facão J, Varga S, Oliveira A C. Evaluation of the use of artificial neural networks for the simulation of hybrid solar collectors[J]. International Journal of Green Energy, 2004, 1(3): 337-352.

[39] Fischer S, Frey P, Drück H. A comparison between state-of-the-art and neural network modelling of solar collectors[J]. Solar Energy, 2012, 86(11): 3268-3277.

[40] Wang S, Zhang T. RETRACTED ARTICLE: Research on innovation path of school ideological and political work based on large data[J]. Cluster Computing, 2019, 22(Suppl 2): 3375-3383.

Chapter 8
Conclusion and Future Work

8.1 Conclusion of the work

To improve the heat transfer performance and efficiency, in this book we propose a novel construction extruded heat exchanger, the ultra-thin superconducting thermal absorber with internal fins shaped bubbling with two different sizes. Furthermore, a PV/T-ASHP system using this kind of absorbers in the solar collector is deployed for an office room real application. Based on this case study, the performance of this system is analyzed through measurements and simulations. The main results of this research are as the following:

For the whole PV/T-ASHP system, our experimental results show that the electricity self-sufficiency reaches its highest value of 78.61%, and in the heating seasons, the thermal self-sufficiency reaches 88.42%, even during December where it can supply 30.73% of the heat demand by the solar collector. The average COP during the heating seasons reaches 4. In general, the ambient temperature and solar irradiation change significantly affect system performance. The larger the difference between the ambient temperature and the outlet water temperature of the solar collector, the higher the COP of ASHP and its electric production efficiency.

Regarding the economic impact, the payback time for this system is 5.6 years, and for renovated buildings, the payback time is 5 years. We further investigate the environmental impact and show that using this system can prevent the emission of 8 t of carbon dioxide, 204.59 kg of sulfur dioxide, and 102.29 kg of nitrogen oxides for the whole year.

It is seen that the environmental factors have a more significant effect on the electric efficiency than that of the operation parameters whereas for the thermal efficiency it is the other way around. Nevertheless, the performance prediction based on all nine factors is required for both electric and thermal efficiency. For this prediction, we show that the ANN with two hidden layers generally performs better than that of one hidden layer for predicting both electric and thermal efficiency. The best ANN network construction for electric efficiency is 9-25-25-1, with the lowest velue of RMSE (0.76%). For thermal efficiency, it is 9-10-10-1, with the lowest value of RMSE (0.805%). Our comparisons suggest that the researchers need to find the best suitable ANN network before directly conducting one construction.

8.2 Future work

There are still some limitations that need to be addressed before the versatile development of solar-assisted ASHP systems to the wider market. The main issues are concluded in the following:

The gaps between a simulation or laboratory testing and real applications need to be further investigated.[1] In the future, we need to: (1) develop a common simulation tool; and (2) measure the system performance under standard testing conditions. For the experimental issues, more experiments should be conducted in different conditions. To make it more popular even for the community using in the future, the life cycle performance should be also analyzed.

The main focus of this book is on the collector design. Further research on the design and construction of the solar collector is required.

For performance prediction, multiple regression and ANN are used and compared to predict the performance of the PV/T in this study. Several factors are considered in our study and future works, more methodologies should be analyzed and compared such as GA, and more factors should be considered for system optimization, e.g., the installation angle of the solar collector and different climate regions.

References

[1] Farhat A A M, Ugursal V I. Greenhouse gas emission intensity factors for marginal electricity generation in Canada[J]. International Journal of Energy Research, 2010, 34(15): 1309-1327.